The Concise Book of the Moving Body

Chris Jarmey

Contributor
Thomas W. Myers

Lotus Publishing
Chichester, England

North Atlantic Books
Berkeley, California

First published in 2006 by
Lotus Publishing
9 Roman Way, Chichester, PO19 3QN and
North Atlantic Books
P O Box 12327
Berkeley, California 94712

Anatomical Drawings Amanda Williams
Text and Cover Design Chris Fulcher
Printed and Bound in the UK by The Bath Press

The Concise Book of the Moving Body is sponsored by the Society for the Study of Native Arts and Sciences, a nonprofit educational corporation whose goals are to develop an educational and crosscultural perspective linking various scientific, social, and artistic fields; to nurture a holistic view of arts, sciences, humanities, and healing; and to publish and distribute literature on the relationship of mind, body, and nature.

Acknowledgements
The publisher would like to thank Thomas W. Myers for the use of figures 8.1, 8.2, 8.5–8.9, 8.12, which have been taken from his book *Anatomy Trains* (Churchill Livingstone).

British Library Cataloguing in Publication Data
A CIP record for this book is available from the British Library
ISBN 1 905367 01 5 (Lotus Publishing)
ISBN 1 55643 623 8 (North Atlantic Books)

Library of Congress Cataloging-in-Publication Data

Jarmey, Chris.
   The concise book of the moving body / by Chris Jarmey ; with Thomas
W. Myers.
      p. ; cm.
   Includes index.
   ISBN 1-905367-01-5 (pbk.) -- ISBN 1-55643-623-8 (North Atlantic
Books : pbk.)
   1. Musculoskeletal system. 2. Human locomotion. 3. Human mechanics.
I. Myers, Thomas W., LMT. II. Title.
   [DNLM: 1. Movement--physiology--Atlases. 2. Musculoskeletal System
--Atlases.   WE 17 J37c 2005]
QM100.J38 2005
612.7--dc22

2005030326

Contents

A Note About Peripheral Nerve Supply

The nervous system comprises:

• The central nervous system (i.e. the brain and spinal cord).
• The peripheral nervous system (including the autonomic nervous system, i.e. all neural structures outside the brain and spinal cord).

The peripheral nervous system consists of 12 pairs of cranial nerves and 31 pairs of spinal nerves (with their subsequent branches). The spinal nerves are numbered according to the level of the spinal cord from which they arise (the level is known as the *spinal segment*).

The relevant peripheral nerve supply is listed with each muscle presented in this book, for those who need to know. However, information about the spinal segment* from which the nerve fibres emanate often differs between the various sources. This is because it is extremely difficult for anatomists to trace the route of an individual nerve fibre through the intertwining maze of other nerve fibres as it passes through its plexus (plexus = a network of nerves: from the Latin word meaning *braid*). Therefore, such information has been derived mainly from empirical clinical observation, rather than through dissection of the body.

In order to give the most accurate information possible, I have duplicated the method devised by Florence Peterson Kendall and Elizabeth Kendall McCreary (*see* resources: Muscles Testing and Function). Kendall & McCreary integrated information from six well-known anatomy reference texts; namely, those written by: Cunningham, deJong, Foerster & Bumke, Gray, Haymaker & Woodhall, and Spalteholz. Following the same procedure, and then cross-matching the results with those of Kendall & McCreary, the following system of emphasising the most important nerve roots for each muscle has been adopted in this book (*see* pp.143–163).

Let us take the supinator muscle as our example, which is supplied by the deep radial nerve, C5, **6**, (7). The relevant spinal segment is indicated by the letter [C] and the numbers [5, **6**, (7)]. Bold numbers [e.g. **6**] indicate that most (at least five) of the sources agree. Numbers that are not bold [e.g. 5] reflect agreement by three of four sources. Numbers not in bold and in parenthesis [e.g. (7)] reflect agreement by two sources only, or if more than two sources specifically regarded it as a very minimal supply. If a spinal segment was mentioned by only one source, it was disregarded. Hence, bold type indicates the major innervation; not bold indicates the minor innervation; and numbers in parenthesis suggest possible or infrequent innervation.

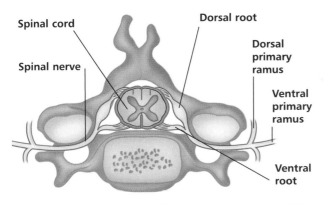

Spinal cord

Spinal nerve

Dorsal root

Dorsal primary ramus

Ventral primary ramus

Ventral root

Figure 1: A spinal segment, showing the nerve roots combining to form a spinal nerve, which then divides into ventral and dorsal rami.

A spinal segment is the part of the spinal cord that gives rise to each pair of spinal nerves (a pair consists of one nerve for the left side and one for the right side of the body). Each spinal nerve contains motor and sensory fibres. Soon after the spinal nerve exits through the foramen (the opening between adjacent vertebrae), it divides into a dorsal primary ramus (directed posteriorly) and a ventral primary ramus (directed laterally or anteriorly) Fibres from the dorsal rami innervate the skin and extensor muscles of the neck and trunk. The ventral rami supply the limbs, plus the sides and front of the trunk.

Anatomical Orientation

1

Anatomical Directions

To describe the relative position of body parts and their movements, it is essential to have a universally accepted initial reference position. The standard body position known as the anatomical position serves as this reference. The *anatomical position* is simply the upright standing position with arms hanging by the sides, palms facing forwards (*see* figure 1.1). Most directional terminology used refers to the body *as if* it were in the anatomical position, regardless of its actual position. Note also that the terms 'left' or 'right' refer to the sides of the object or person being viewed, and not those of the reader.

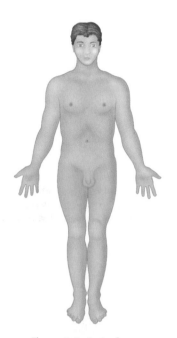

Figure 1.1: **Anterior**
In front of; toward or at the front of the body.

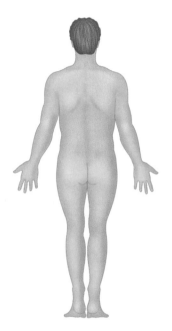

Figure 1.2: **Posterior**
Behind; toward or at the backside of the body.

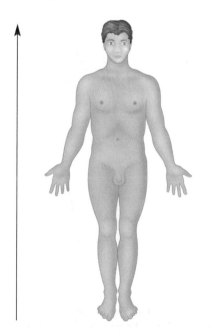

Figure 1.3: **Superior**
Above; toward the head or upper part
of the structure or the body.

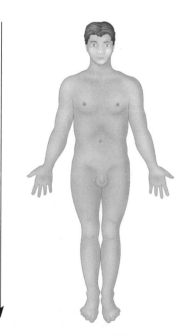

Figure 1.4: **Inferior**
Below; away from the head or toward the
lower part of a structure or the body.

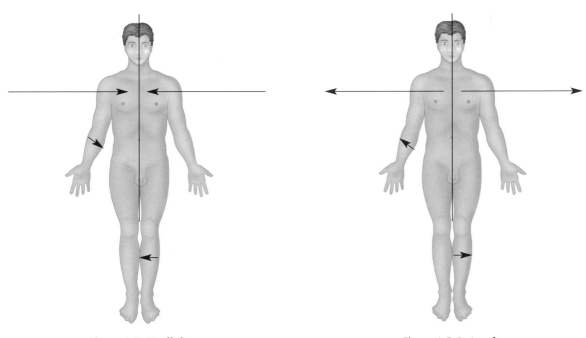

Figure 1.5: **Medial**
(from *medius* in Latin, meaning middle). Toward or at the midline of the body; on the inner side of a limb.

Figure 1.6: **Lateral**
(from *latus* in Latin, meaning side). Away from the midline of the body; on the outer side of the body or a limb.

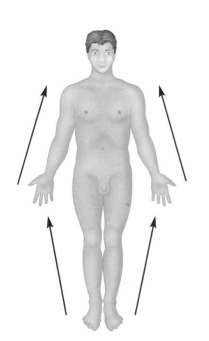

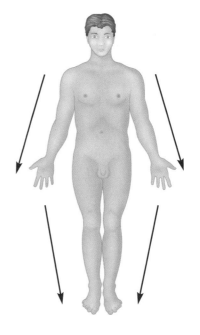

Figure 1.7: **Proximal**
(from *proximus* in Latin, meaning next to). Closer to the centre of the body (the navel), or to the point of attachment of a limb to the body torso.

Figure 1.8: **Distal**
(from *distans* in Latin, meaning distant). Farther from the centre of the body, or from the point of attachment of a limb to the torso.

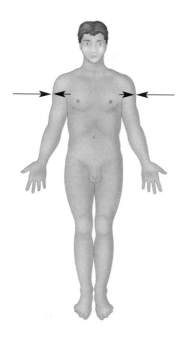

Figure 1.9: **Superficial**
Toward or at the body surface.

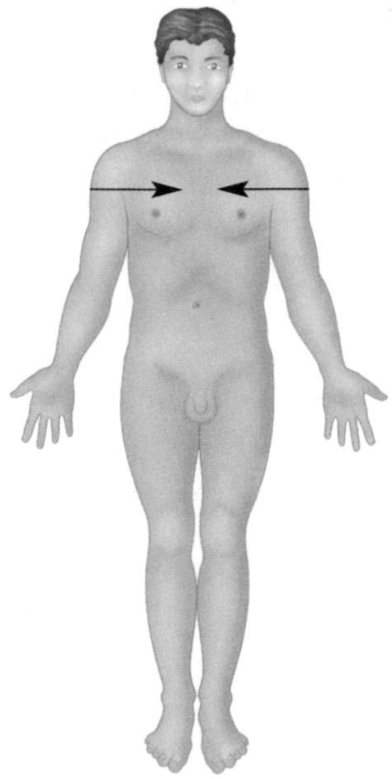

Figure 1.10: **Deep**
Farther away from the body surface; more internal.

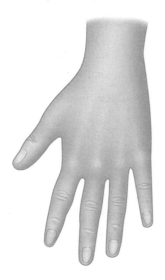

Figure 1.11: **Dorsum**
The posterior surface of something,
e.g. the back of the hand;
the top of the foot.

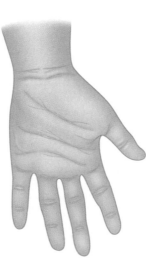

Figure 1.12: **Palmar**
The anterior surface of the hand,
i.e. the palm.

Figure 1.13: **Plantar**
The sole of the foot.

Regional Areas

The two primary divisions of the body are its *axial* part, consisting of the head, neck and trunk, and its *appendicular* parts, consisting of the limbs that are attached to the axis of the body. Figure 1.14 shows the terms used to indicate specific body areas. Terms enclosed within brackets refer to the lay term for the area.

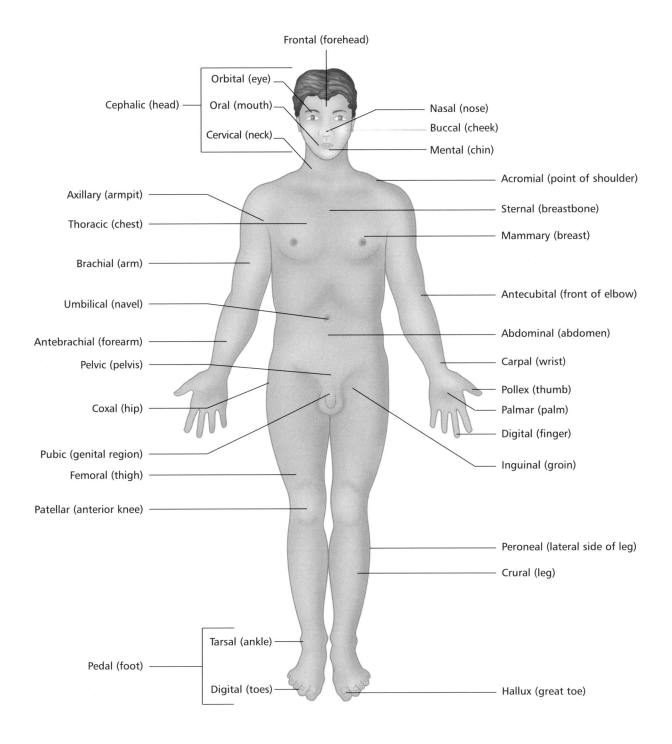

Figure 1.14: Terms used to indicate specific body areas; a) anterior view.

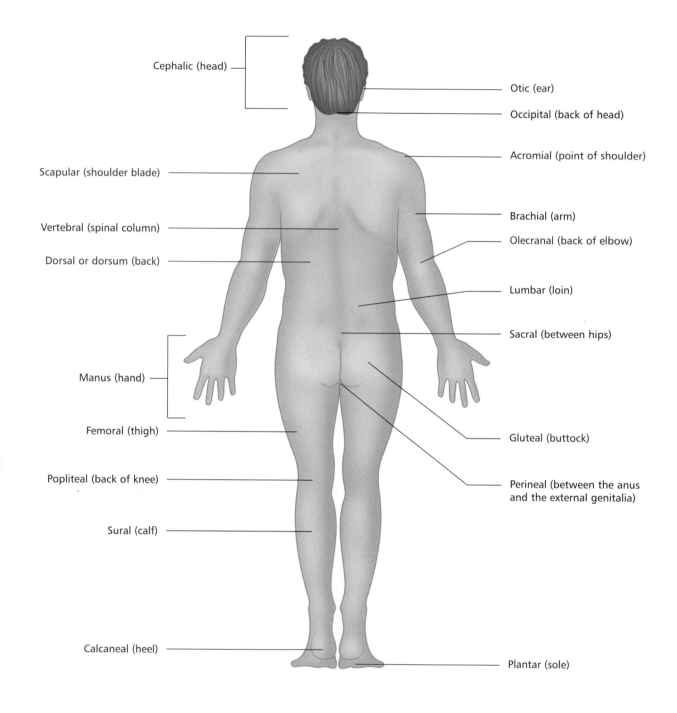

Cephalic (head)

Scapular (shoulder blade)

Vertebral (spinal column)

Dorsal or dorsum (back)

Manus (hand)

Femoral (thigh)

Popliteal (back of knee)

Sural (calf)

Calcaneal (heel)

Otic (ear)

Occipital (back of head)

Acromial (point of shoulder)

Brachial (arm)

Olecranal (back of elbow)

Lumbar (loin)

Sacral (between hips)

Gluteal (buttock)

Perineal (between the anus and the external genitalia)

Plantar (sole)

Figure 1.14: Terms used to indicate specific body areas; b) posterior view.

Planes of the Body

Planes refer to two-dimensional sections through the body, to give a view of the body or body part, as if it has been cut through an imaginary line.

• The sagittal planes cut vertically through the body from anterior to posterior, dividing the body into right and left halves. The illustration shows the mid-sagittal plane.

• The frontal (coronal) planes pass vertically through the body, dividing the body into anterior and posterior sections, and lies at right angles to the sagittal plane.

• The transverse planes are horizontal cross sections, dividing the body into upper (superior) and lower (inferior) sections, and lie at right angles to the other two planes. Figure 1.15 illustrates the most frequently used planes.

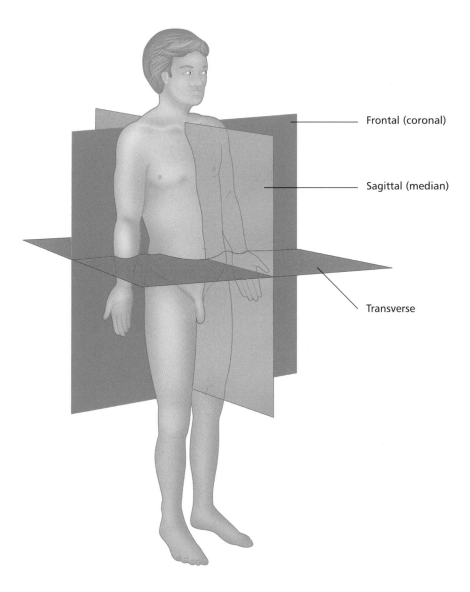

Frontal (coronal)

Sagittal (median)

Transverse

Figure 1.15: Planes of the body.

Anatomical Movements

The direction that body parts move is described in relation to the foetal (fetal) position. Moving into the foetal position results from flexion of all the limbs. Straightening out of the foetal position results from extension of all the limbs.

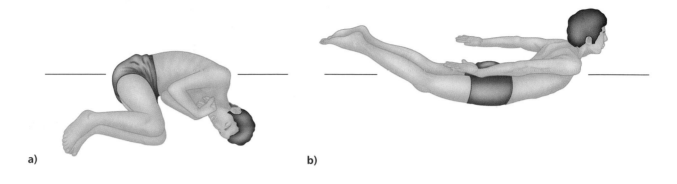

a)   b)

Figure 1.16: a) Flexion into the foetal position; b) extension out of the foetal position.

Main Movements

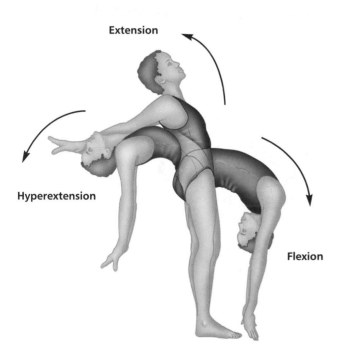

Figure 1.17: **Flexion**: Bending to decrease the angle between bones at a joint. From the anatomical position, flexion is usually forward, except at the knee joint where it is backward. The way to remember this is that flexion is always toward the foetal position. **Extension**: To straighten or bend backward away from the foetal position. **Hyperextension**: To extend the limb beyond its normal range.

Figure 1.18: **Lateral flexion**
To bend the torso or head laterally (sideways) in the frontal (coronal) plane.

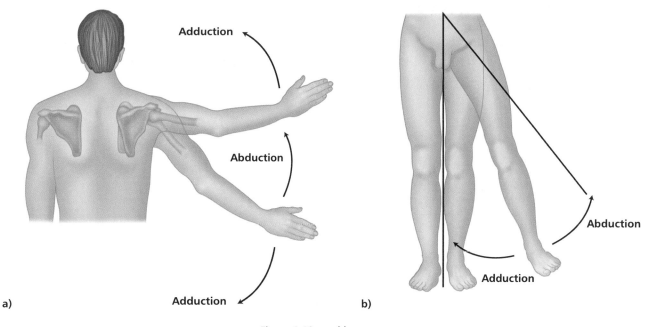

Figure 1.19a and b:
Abduction: Movement of a bone away from the midline of the body, or the midline of a limb.
Adduction: Movement of a bone towards the midline of the body, or the midline of a limb.

NOTE: for abduction of the arm to continue above the height of the shoulder (elevation through abduction), the scapula must rotate on its axis to turn the glenoid cavity upwards (*see* figure 1.27b).

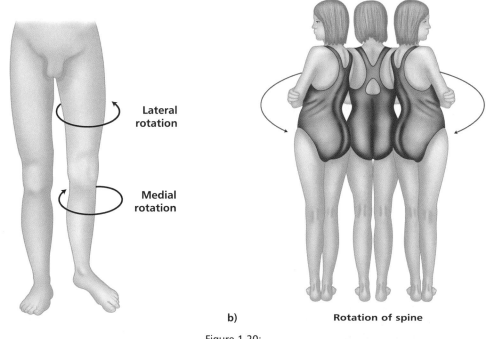

Figure 1.20:
Rotation: Movement of a bone or the trunk around its own longitudinal axis.
Medial rotation: to turn in towards the midline.
Lateral rotation: to turn out, away from the midline.

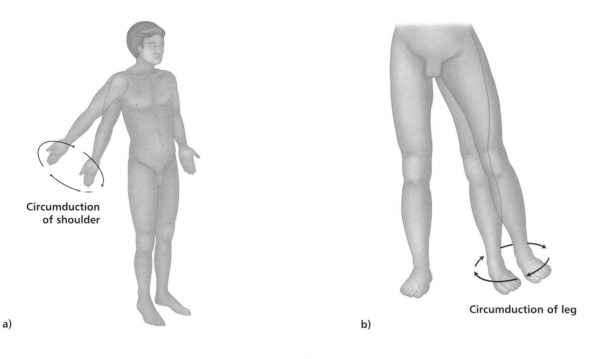

a)   b)

Figure 1.21: **Circumduction**
Movement in which the distal end of a bone moves in a circle, while the proximal end remains stable;
the movement combines flexion, abduction, extension, and adduction.

Other Movements

Movements in this section are those that occur only at specific joints or parts of the body; usually involving more than one joint.

a)

b)

Figure 1.22a: **Pronation**
To turn the palm of the hand down to face the floor
(if standing with elbow bent 90°, or if lying flat on the
floor), or away from the anatomical and foetal positions.

Figure 1.22b: **Supination**
To turn the palm of the hand up to face the ceiling
(if standing with elbow bent 90°, or if lying flat on the
floor), or toward the anatomical and foetal positions.

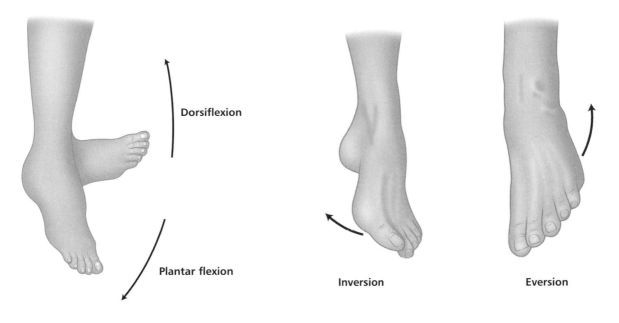

Figure 1.23: **Plantar flexion**: To point the toes down towards the ground. **Dorsiflexion**: To point the toes up towards the sky.

Figure 1.24: **Inversion**: To turn the sole of the foot inward, so that the soles would face towards each other. **Eversion**: To turn the sole of the foot outward, so that the soles would face away from each other.

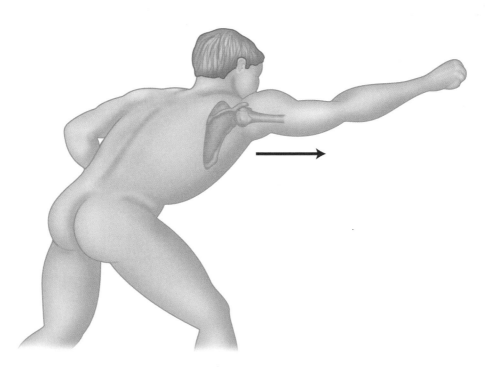

Figure 1.25: **Protraction**
Movement forwards in the transverse plane.
For example, protraction of the shoulder girdle, as in rounding the shoulder.

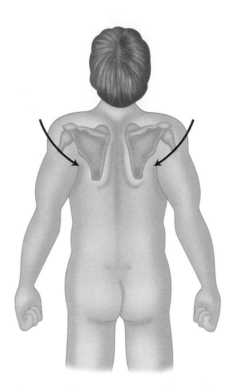

Figure 1.26: **Retraction**
Movement backwards in the transverse plane,
as in bracing the shoulder girdle back, military style.

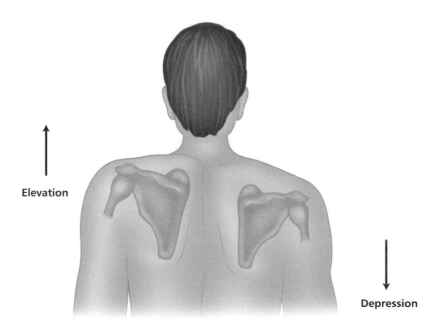

Elevation

Depression

Figure 1.27a:
Elevation: Movement of a part of the body upwards along the frontal plane.
For example, elevating the scapula by shrugging the shoulders.
Depression: Movement of an elevated part of the body downwards to its original position.

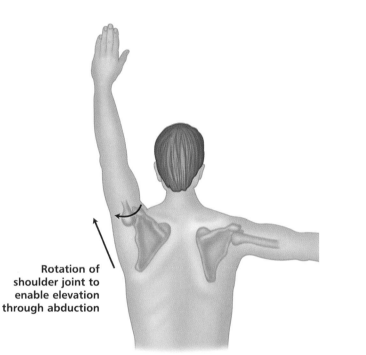

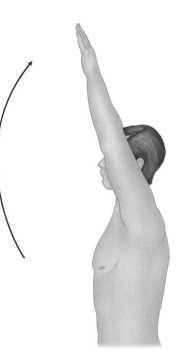

Rotation of shoulder joint to enable elevation through abduction

Figure 1.27b: Abducting the arm at the shoulder joint, then continuing to raise it above the head in the frontal plane can be referred to as **elevation through abduction**.

Figure 1.27c: Flexing the arm at the shoulder joint, then continuing to raise it above the head in the sagittal plane can be referred to as **elevation through flexion**.

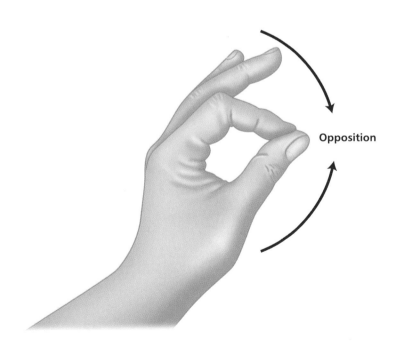

Opposition

Figure 1.28: **Opposition**
A movement specific to the saddle joint of the thumb,
that enables you to touch your thumb to the tips of the fingers of the same hand.

Tissues

2

Connective Tissue

Muscle Tissue

Groups of cells having a similar structure and function are called tissues. There are four primary types of tissue, each having a characteristic function:

1. Epithelial
For example, the skin, whose function is to cover and protect the body.

2. Connective
For example, ligaments, tendons, fascia, cartilage, and bone, whose functions are to protect, support, and bind together different parts of the body. It is the most abundant and widely distributed tissue type.

3. Muscle
The function of muscle is to enable movement of the body.

4. Nervous
Nerves control body functions and movement.

The types of tissue that concern us regarding musculo-skeletal anatomy are: *connective tissue* and *muscle tissue*, for which a brief overview is given below:

Connective Tissue

Common Characteristics
1. Variations in blood supply – Most connective tissue is well vascularized (having blood vessels), but tendons and ligaments have a poor blood supply, and cartilage is avascular (having no blood vessels), therefore these structures heal very slowly.

2. Extracellular matrix – Connective tissues are made up of many different types of cells and varying amounts of nonliving substance that surrounds the cells. This substance is called *extracellular matrix*. This matrix is produced by connective tissue cells and then secreted to their exterior. Depending on the type of tissue, the matrix may be: liquid, gel-like, semisolid, or very hard.

Because of the matrix, connective tissue is able to bear weight, and to withstand stretching and other abuses, such as abrasion. The matrix contains various types and amounts of fibres, e.g. collagen, elastic, or reticular.

a) 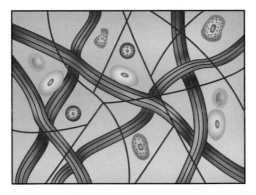   b)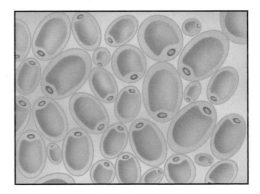

Figure 2.1: Structure of loose connective tissue; a) areolar, b) adipose.

Types of Connective Tissue

Loose Connective Tissue

Loose connective tissue has more cells and fewer fibres; thus, it is softer than the other types. Examples are:

1. *Areolar*, a 'packing' tissue, which cushions and protects body organs and holds internal organs together in their proper position;
2. *Adipose*, fat tissue that forms the subcutaneous layer beneath the skin, also called the *hypodermis*, or *superficial fascia*, where it insulates the body and protects against heat and cold.

Dense Regular Connective Tissue

Within dense regular connective tissue, collagen fibres are the predominant element and create a white, flexible tissue with great resistance to pulling forces. Examples are: *ligaments* and *tendons*.

Dense Irregular Connective Tissue

Dense irregular connective tissue has the same structural elements as regular connective tissue. However the bundles of collagen fibres are thicker, are interwoven and are arranged irregularly. Fascia is an example of dense irregular connective tissue.

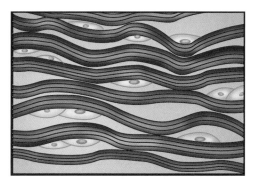
Figure 2.2: Structure of dense regular connective tissue.

Figure 2.3: Structure of dense irregular connective tissue.

Cartilage

Cartilage is tough, but flexible. It has qualities intermediate between dense connective tissue and bone. Cartilage is avascular and devoid of nerve fibres, and therefore heals slowly. Examples are: *hyaline*, *fibrocartilage*, and *elastic*.

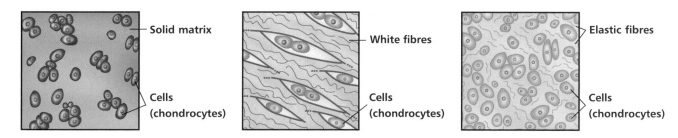
Figure 2.4: Structure of cartilage; a) hyaline cartilage, b) white fibrocartilage, c) yellow elastic cartilage.

Bone

Bone cells sit in cavities called *lacunae* (sing. *lacuna*) surrounded by circular layers of a very hard matrix that contains calcium salts and larger amounts of collagen fibres.

Blood

Blood, or vascular tissue, is considered a connective tissue because it consists of blood cells, surrounded by a nonliving fluid matrix called *blood plasma*. The 'fibres' of blood are soluble protein molecules that become visible only during blood clotting. Blood is not a typical connective tissue; it is the transport vehicle for the cardiovascular system, carrying nutrients, wastes, respiratory gases, etc. throughout the body.

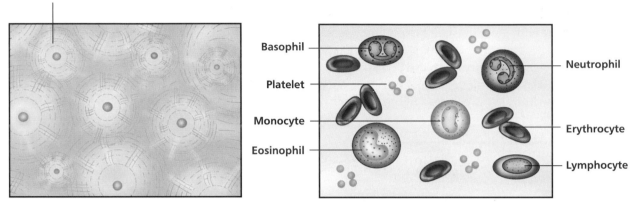

Figure 2.5: Structure of bone.   Figure 2.6: Structure of blood.

Muscle Tissue

Muscle is composed of 75% water, 20% protein, and 5% mineral salts, glycogen and fat. As this book is designed to focus specifically on musculo-skeletal anatomy, only a brief description and comparison of the different types of muscle tissue is given below. Skeletal muscle is then discussed in more detail (*see* Chapter 7).

Muscle Types and Function

There are three types of muscle tissue: skeletal, cardiac and smooth. All muscle cells have an elongated shape and are therefore referred to as *muscle fibres*.

Smooth / Unstriated / Involuntary Muscle

Smooth muscle cells are usually spindle-shaped and arranged in sheets or layers. Smooth muscles are found in the *viscera*, i.e. stomach, small and large intestines, blood vessels, uterus (i.e. the hollow organs).

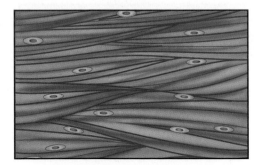

Figure 2.7: Structure of smooth / unstriated / involuntary muscle.

Smooth muscle in the blood vessels contracts to move the blood in the arteries. Smooth muscle also squeezes substances through the organs and tracts. They are under *involuntary* control (although some individuals can train their minds to achieve some control over smooth muscle contractions). Contractions are usually gentle and rhythmic, with the obvious exceptions of vomiting and birth contractions.

Cardiac / Striated / Involuntary Muscle
Cardiac muscles are found in the heart only, and exist to pump the heart. They are under *involuntary* control. Structurally, they are made up of branching fibres that are striated in appearance and are separated or interspersed by discs, known as *intercalated discs*.

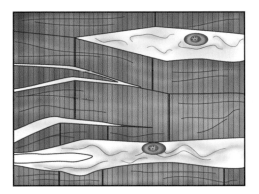

Figure 2.8: Structure of cardiac / striated / involuntary muscle.

Skeletal / Striated / Voluntary Muscle
Skeletal muscles (also called somatic muscles) attach to, and cover over, the bony skeleton. They are under *voluntary* control. Skeletal muscles fatigue easily, but can be strengthened. They are capable of powerful, rapid contractions, and longer, sustained contractions. Skeletal muscles enable us to perform both feats of strength and controlled, fine movements.

NOTE: As they contract, all muscle types generate heat, and this heat is vitally important in maintaining a normal body temperature. It is estimated that 85% of all body heat is generated by muscle contractions.

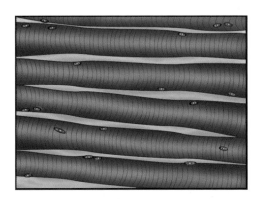

Figure 2.9: Structure of skeletal / striated / voluntary muscle.

Bone

3

We are born with approximately 350 bones, but gradually they fuse together until by puberty we have only 206 bones. These bones form the supporting structure of the body, and are collectively known as the endoskeleton. (The exoskeleton is well developed in many invertebrates, but exists in humans only as teeth, nails and hair). Fully developed bone is the hardest tissue in the body and is composed of 20% water, 30% to 40% organic matter, and 40% to 50% inorganic matter.

Bone Development and Growth

The majority of bone is formed from a foundation of cartilage (*see* below), which becomes calcified and then ossified to form true bone. This process occurs through the following four stages:

1. Bone building cells called *osteoblasts* become active during the second or third month of embryonic life.
2. Initially, the osteoblasts manufacture a *matrix* of material between the cells, which is rich in a fibrous protein called *collagen*. This collagen strengthens the tissue. Enzymes then enable calcium compounds to be deposited within the matrix.
3. This intercellular material hardens around the cells, to become *osteocytes*; i.e. living cells that maintain the bone, but do not produce new bone.
4. Other cells, called *osteoclasts*, breakdown, remodel and repair bone; a process that continues throughout life, but which slows down with advancing age. Consequently, the bones of elderly people are weaker and more fragile.

In brief, osteoblasts and osteoclasts are the cells that lay down and break down bone respectively, enabling bones to very slowly adapt in shape and strength according to need.

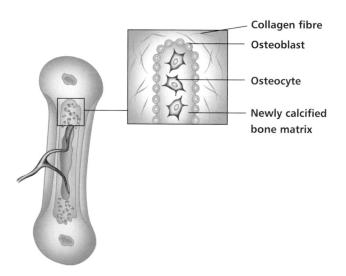

Figure 3.1: Bone development and growth.

Cartilage

Cartilage (gristle) exists either as a temporary formation that is later replaced by bone, or as a permanent supplementation to bone. However, it is not as hard or as strong as bone.

It consists of living cells called *chondrocytes*, contained within *lacunae* (spaces) and surrounded by a collagen rich intercellular substance. Cartilage is relatively non-vascular (not penetrated by blood vessels) and is mainly nourished by surrounding tissue fluids. There are three main types of cartilage: hyaline cartilage, white fibrocartilage and yellow fibrocartilage.

Hyaline Cartilage

Hyaline cartilage forms the temporary foundation of cartilage from which many bones develop, thereafter existing in relation to bone as:

• The articular cartilage of synovial joints.
• Cartilage plates between separately ossifying areas of bone during growth.
• The xiphoid process of the sternum (which ossifies late or not at all) and the costal cartilages.

Hyaline cartilage also exists in the nasal septum, most cartilages of the larynx, and the supporting rings of the trachea and bronchi.

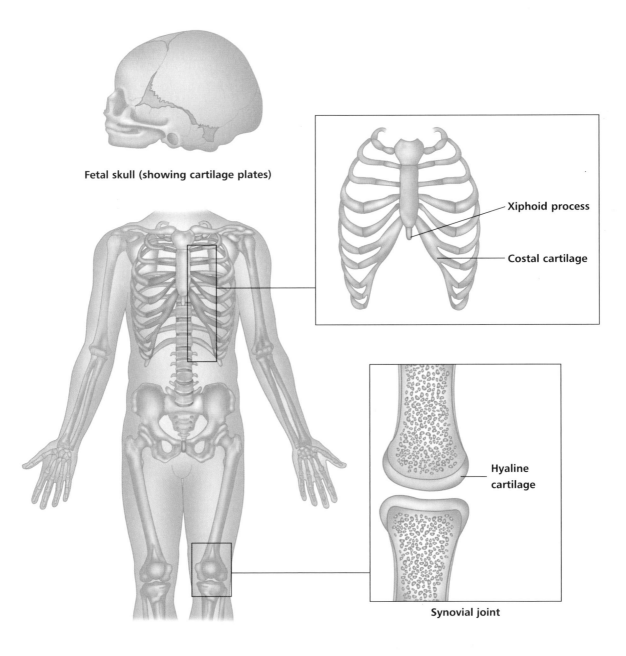

Fetal skull (showing cartilage plates)

Xiphoid process

Costal cartilage

Hyaline cartilage

Synovial joint

Figure 3.2: Location sites of hyaline cartilage in the body.

White Fibrocartilage

White fibrocartilage contains white fibrous tissue. It has more elasticity and tensile strength than hyaline cartilage. It is found as the:

- Sesamoid cartilages in a few tendons.
- Articular discs in the wrist joint and clavicular joints.
- Rim (labrum) deepening the sockets of the shoulder and hip joints.
- Two semilunar cartilages within each knee joint.
- Intervertebral discs between adjacent surfaces of the vertebral bodies.
- Cartilage plate joining the hipbones at the pubic symphysis.

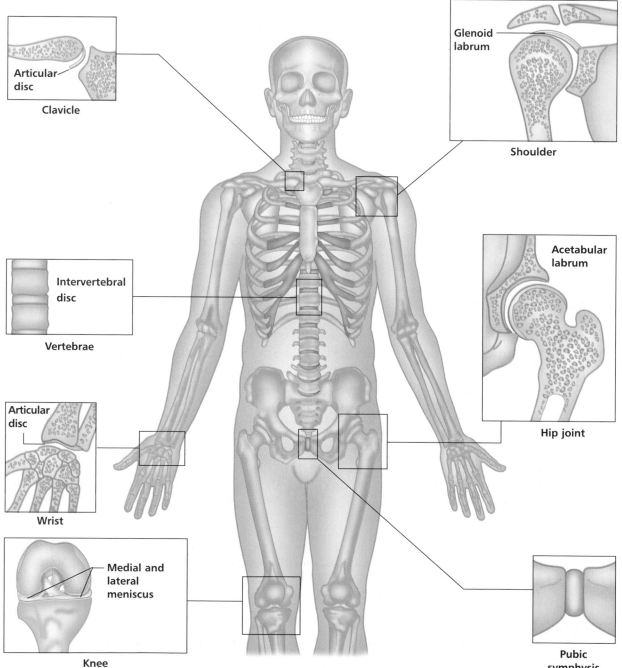

Figure 3.3: Location sites of white fibrocartilage in the body.

Yellow Fibrocartilage

Yellow fibrocartilage contains yellow elastic fibres. It is found in the external ear, auditory tube of the middle ear, and the epiglottis.

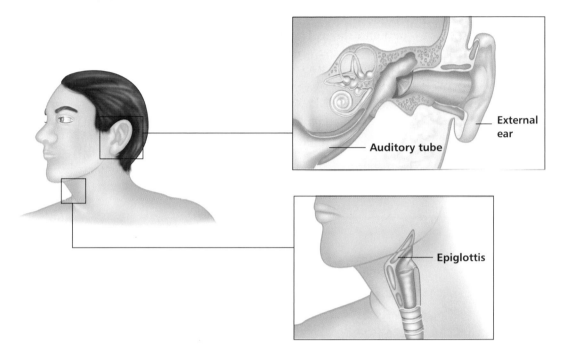

Figure 3.4: Location sites of yellow fibrocartilage in the body.

Functions of Bones

Support

Our bones provide the hard framework that supports and anchors all the soft organs of the body. Our legs support the body's torso, head, and arms. The ribcage supports the chest wall.

Protection

The bones of the skull protect the brain; the vertebrae surround the spinal cord; the ribcage protects all the vital organs.

Movement

Muscles are attached to the bones by tendons; and they use the bones as levers to move the body and all its parts; the arrangement of the bones and joints determines which movements are possible.

Storage

Fat is stored as 'yellow marrow' in the central cavities of long bones. Within the structure of the bone itself minerals are stored. The most important minerals are calcium and phosphorus, but potassium, sodium, sulphur, magnesium, and copper are also stored. Stored minerals can be released into the bloodstream for distribution to all parts of the body as needed.

Blood Cell Formation

The bulk of blood cell formation occurs within the 'red marrow' cavities of certain bones.

Types of Bone – According to Density

Compact Bone

Compact bone is dense, and looks smooth to the naked eye. Through the microscope, compact bone appears as an aggregation of *Haversian systems*, also called *osteons*. Each such system is an elongated cylinder oriented along the long axis of the bone, consisting of a central *Haversian canal* containing blood vessels, lymph vessels and nerves, surrounded by concentric plates of bone called *lamellae*. In other words, each Haversian system is a group of hollow tubes of bone matrix (lamellae), placed one inside the next. Between these lamellae there are spaces (*lacunae*) that contain lymph and osteocytes. The lacunae are linked via hair-like canals called *canaliculi* to the lymph vessels in the Haversian canal, enabling the osteocytes to obtain nourishment from the lymph. This tubular array of lamellae gives great strength to the bone.

Other canals called *perforating*, or *Volkmann's canals*, run at right angles to the long axis of the bone, connecting the blood vessels and nerve supply within the bone to the periosteum (*see* page 35).

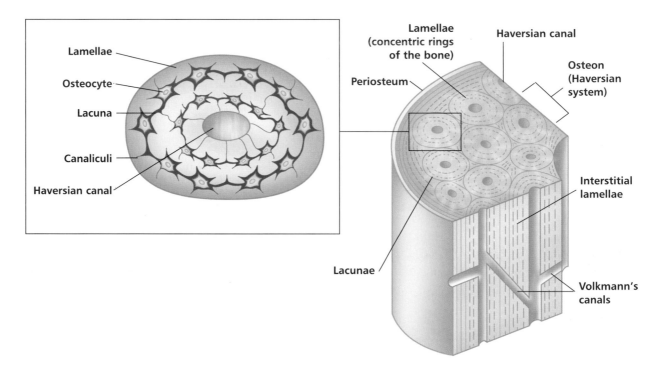

Figure 3.5: Structure of compact bone.

Spongy Bone (Cancellous Bone)

Spongy bone is composed of small needle-like *trabeculae* (sing. *trabecula*; literally, *little beam(s)*) containing irregularly arranged lamellae and osteocytes, interconnected by canaliculi. There are no Haversian systems, but rather, lots of open spaces, that can be thought of as large Haversian canals, giving a honeycombed appearance. These spaces are filled with red or yellow marrow and blood vessels.

This structure forms a dynamic lattice capable of gradual alteration through realignment, in response to stresses of weight, postural change, and muscle tension. Spongy bone is found in the epiphyses of long bones, the bodies of the vertebrae, and other bones without cavities.

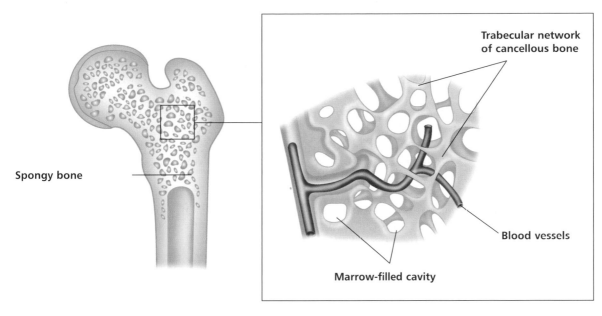

Figure 3.6: Structure of spongy (cancellous) bone.

Types of Bones – According to Shape

Irregular Bones

Irregular bones have complicated shapes; they consist mainly of spongy bone enclosed by thin layers of compact bone. Examples include: some skull bones, the vertebrae, and the hipbones.

Flat Bones

Flat bones are thin, flattened bones, and frequently curved; they have a layer of spongy bone sandwiched between two thin layers of compact bone. Examples include: most of the skull bones, the ribs, and the sternum.

Short Bones

Short bones are generally cube-shaped; consist mostly of *spongy (cancellous)* bone. Examples include: the carpal bones in the hand, and tarsal bones in the ankle.

– **sesamoid** (from the Latin, meaning *shaped like a sesame seed*): Sesamoid bones are a special type of short bone that are formed and embedded within a tendon. Examples are: the patella (kneecap) and the pisiform bone at the medial end of the wrist crease.

Long Bones

Long bones are longer than they are wide; they have a shaft with heads at both ends, and consist mostly of compact bone. Examples include: the bones of the limbs, except those of the wrist, hand, ankle and foot (although the bones of the fingers and toes are effectively miniature long bones).

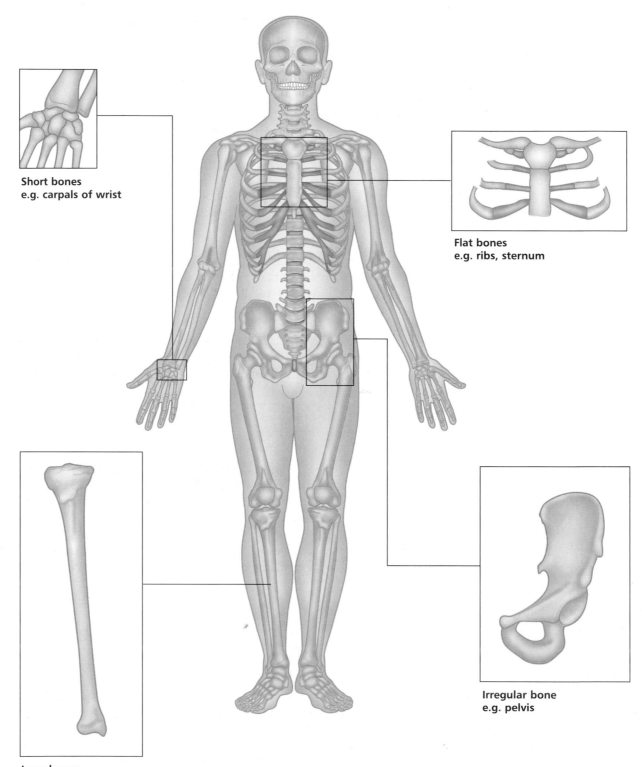

Short bones
e.g. carpals of wrist

Flat bones
e.g. ribs, sternum

Long bones
e.g. tibia

Irregular bone
e.g. pelvis

Figure 3.7: Bone shapes.

Components of a Long Bone

The transformation of cartilage within a long bone begins at the centre of the shaft. Secondary bone-forming centres develop later on, across the ends of the bones. From these growth centres, the bone continues to grow through childhood and adolescence, finally ceasing in the early twenties, whereupon the growth regions harden.

Diaphysis (from Greek, meaning *a separation*)
The diaphysis is the shaft or central part of a long bone. It has a marrow filled cavity (medullary cavity) surrounded by compact bone. It is formed from one or more primary sites of ossification, and supplied by one or more nutrient arteries.

Epiphysis (from Greek, meaning *excrescence*)
The epiphysis is the end of a long bone, or any part of a bone separated from the main body of an immature bone by cartilage. It is formed from a secondary site of ossification. It consists largely of *spongy bone*.

Epiphyseal Line
The epiphyseal line is the remnant of the epiphyseal plate (a flat plate of hyaline cartilage) seen in young, growing bone. It is the site of growth of a long bone. By the end of puberty, long bone growth stops and this plate is completely replaced by bone, leaving just the line to mark its previous location.

Articular Cartilage
Articular cartilage is the only remaining evidence of an adult bone's cartilaginous past. It is located where two bones meet (articulate) within a synovial joint. It is smooth, slippery, porous, malleable, insensitive, and bloodless. It is massaged by movement, which permits absorption of synovial fluid, oxygen, and nutrition.

NOTE: The degenerative process of osteoarthritis (and the latter stages of some forms of rheumatoid arthritis) involves the breakdown of articular cartilage.

Periosteum
The periosteum is a fibrous connective tissue membrane that is vascular and provides a highly sensitive double-layered life support sheath enveloping the outer surface of bone. The outer layer is made of dense irregular connective tissue. The inner layer, which lies directly against the bone surface, mostly comprises the bone-forming *osteoblasts* and the bone-destroying *osteoclasts*.

The periosteum is supplied with nerve fibres, lymphatic vessels, and blood vessels that enter the bone through *nutrient canals*. It is attached to the bone by collagen fibres, known as *Sharpey's fibres*. The periosteum also provides the anchoring point for tendons and ligaments.

Medullary Cavity
The medullary cavity is the cavity of the diaphysis (i.e. the central section of a long bone). It contains marrow: red in the young, turning to yellow in many bones in maturity.

Red Marrow
Red marrow is a red, gelatinous substance composed of red and white blood cells in a variety of developmental forms. The *red marrow cavities* are typically found within the spongy bone of long bones and flat bones. In adults the red marrow, which creates new red blood cells, occurs only in the head of the femur and the head of the humerus, and, much more importantly, in the flat bones such as the sternum and irregular bones, such as the hipbones. These are the sites routinely used for obtaining red marrow samples when problems with the blood-forming tissues are suspected.

Yellow Marrow
Yellow marrow is a fatty connective tissue that no longer produces blood cells.

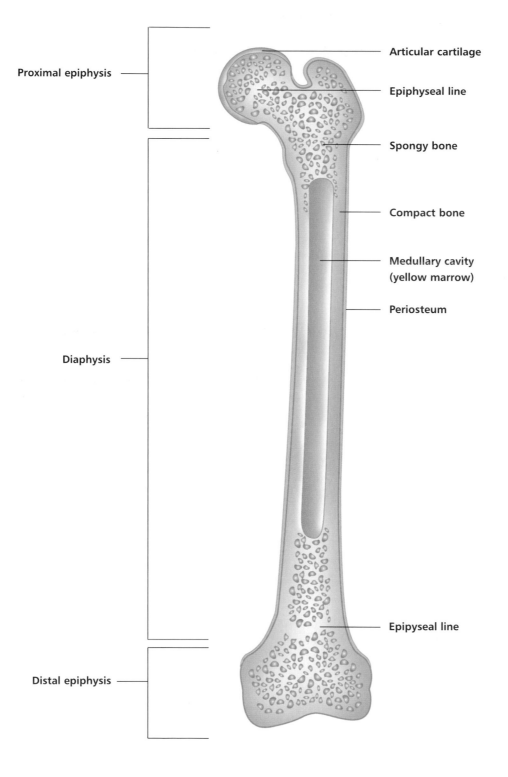

Proximal epiphysis

Diaphysis

Distal epiphysis

Articular cartilage

Epiphyseal line

Spongy bone

Compact bone

Medullary cavity
(yellow marrow)

Periosteum

Epipyseal line

Figure 3.8: Components of a long bone.

Bone Markings

Bone markings fall into three broad categories, as given below:

1. Projections on Bones that are the Sites of Muscle and Ligament Attachment

Trochanter
A very large, blunt, and irregularly shaped projection. The only example is on the femur.

Tuberosity
A large rounded projection, which may be roughened. The main examples are on the tibia (tibial tuberosities) and the ischium (ischial tuberosities).

Tubercle
A smaller rounded projection, which may be roughened.

Crest
A projection, or projecting, narrow ridge of bone. Usually prominent, notably the iliac crest.

Border
A bounding line or edge. Also called margin. Narrow ridge of bone that separates two surfaces.

Spine or Spinous Process
A sharp, slender, often pointed projection; notably the spinous processes on the vertebrae; and the spines of the scapula or the ilium (anterior superior iliac spine, abbreviated as the ASIS, and the posterior superior iliac spine, abbreviated as the PSIS).

Epicondyle
A raised area, on or above a condyle. Notably on the humerus at the elbow joint.

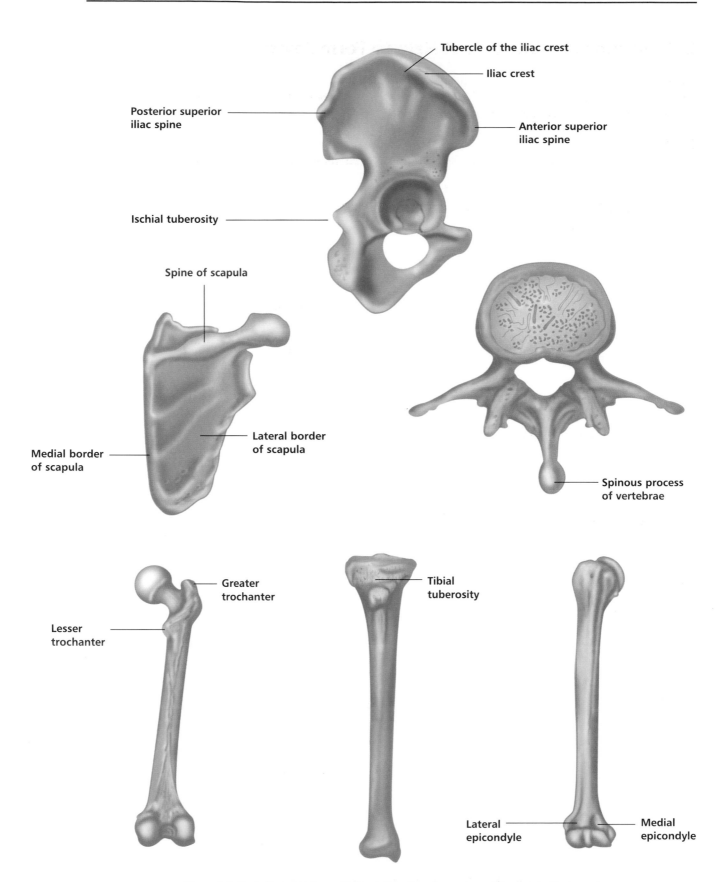

Figure 3.9: Projections on bones that are the sites of muscle and ligament attachment.

2. Projections on Bones that Help to Form Joints

Head
An expansion, which is usually round, located at one end of a bone. For example the head of the fibula, which articulates with the tibia, just below the knee joint.

Facet
A smooth, nearly flat surface at one end of bone, which articulates with another bone.

Condyle
A large rounded projection, which articulates with another bone (found at the knee joint).

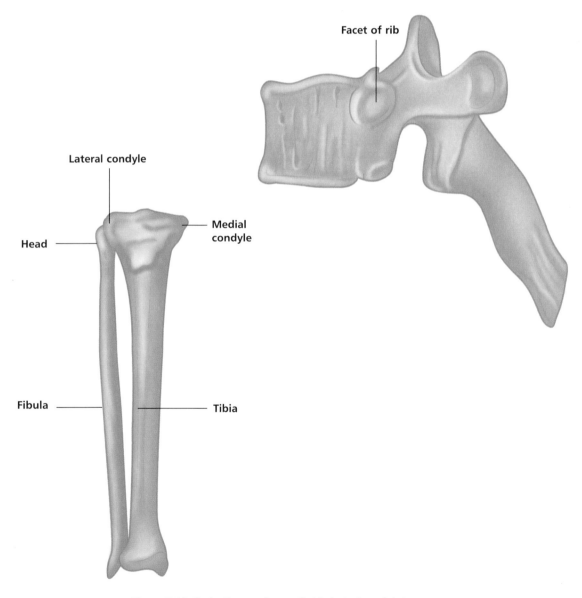

Figure 3.10: Projections on bones that help to form joints.

3. Depressions and Openings that Allow Blood Vessels and Nerves to Pass Through

Sinus
A cavity within a bone, that is filled with air and lined with a membrane (most notably in the skull).

Fossa
A shallow, basin-like depression in a bone. Often serving as an articular surface.

Foramen (pl. *foramina*)
A round or oval opening through a bone (most notably on the sacrum).

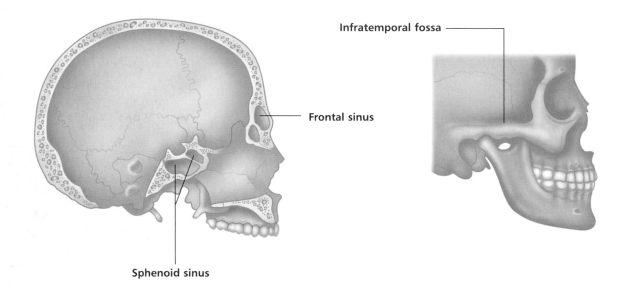

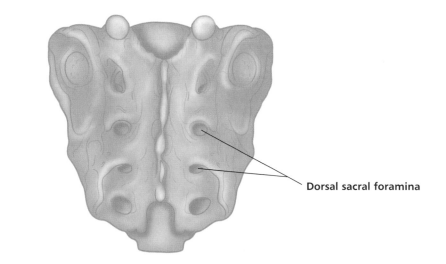

Figure 3.11: Depressions and openings that allow blood vessels and nerves to pass through.

The Axial
Skeleton

4

The Skull:
Comprising the
Cranium and the
Facial Bones

The Vertebral
Column (Spine)

The Bony Thorax

The Skull: Comprising the Cranium and the Facial Bones

NOTE: Some soft tissues (e.g. cartilages, aponeuroses, ligaments and tendons) are included where appropriate, for ease of reference.

The Cranium

The cranium (Greek: *kranion*, the upper part of the head) is made up of eight large flat bones: comprising two pairs, plus four single bones. These form a box-like container that houses the brain. These bones are:

Frontal: which forms the forehead, the bony projections under the eyebrows and the superior part of each eye orbit.

Parietal: a pair of bones that form most of the superior and lateral walls of the cranium. They meet in the midline at the *sagittal suture*, and meet with the frontal bone at the *coronal suture*.

Temporal: a pair of bones which lie inferior to the parietal bones; there are three important markings on the temporal bone: (a) the *styloid process* which is just in front of the mastoid process; a sharp needle-like projection to which many of the neck muscles attach; (b) the *zygomatic process*, a thin bridge of bone that joins with the zygomatic bone just above the mandible; (c) the *mastoid process*, a rough projection posterior and inferior to the styloid process (just behind the lobe of the ear).

Occipital: the most posterior bone of the cranium. It forms the floor and back wall of the skull; and joins the parietal bones anteriorly at the *lambdoidal suture*. In the base of the occipital bone is a large opening, the *foramen magnum* through which the spinal cord passes to connect with the brain. To each side of the foramen magnum are the *occipital condyles* that rest on the first vertebra of the spinal column (the atlas).

Sphenoid: a butterfly-shaped bone that spans the width of the skull and forms part of the floor of the cranial cavity. Parts of the sphenoid can be seen forming part of the eye orbits, and the lateral part of the skull.

Ethmoid (*see* figure 4.3): a single bone in front of the sphenoid bone and below the frontal bone. Forms part of nasal septum and superior and medial conchae.

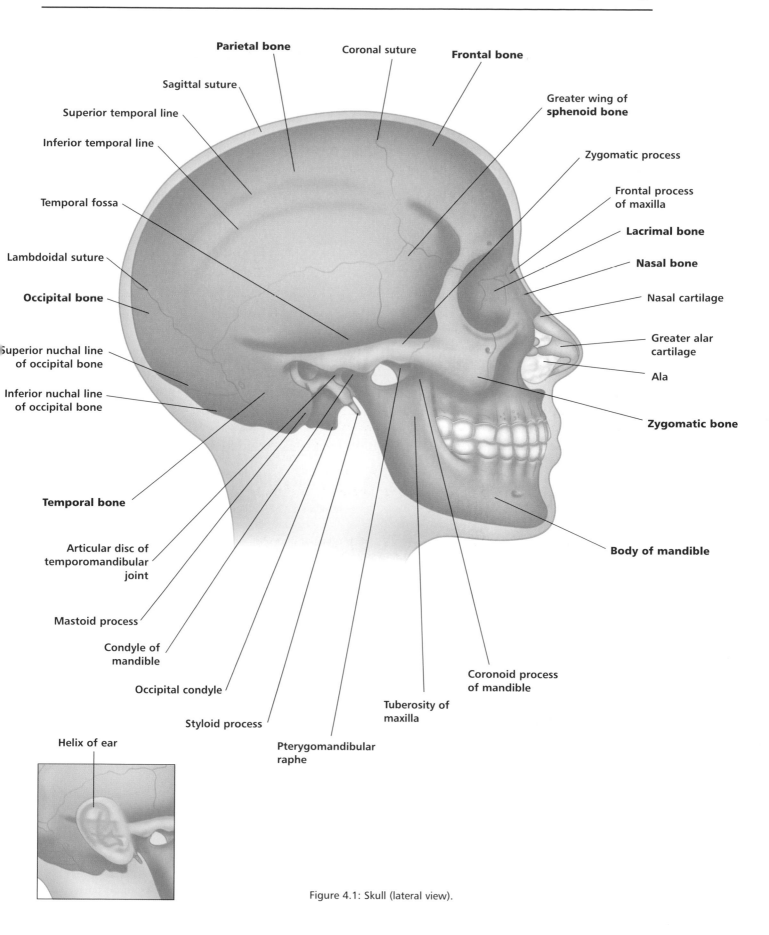

Parietal bone

Coronal suture

Frontal bone

Sagittal suture

Superior temporal line

Inferior temporal line

Temporal fossa

Lambdoidal suture

Occipital bone

Superior nuchal line
of occipital bone

Inferior nuchal line
of occipital bone

Temporal bone

Articular disc of
temporomandibular
joint

Mastoid process

Condyle of
mandible

Occipital condyle

Styloid process

Pterygomandibular
raphe

Helix of ear

Greater wing of
sphenoid bone

Zygomatic process

Frontal process
of maxilla

Lacrimal bone

Nasal bone

Nasal cartilage

Greater alar
cartilage

Ala

Zygomatic bone

Body of mandible

Coronoid process
of mandible

Tuberosity of
maxilla

Figure 4.1: Skull (lateral view).

The Facial Bones

Fourteen bones compose the face, twelve of which are pairs. The main bones of the face are:

Nasal: a pair of small rectangular bones that form the bridge of the nose (the lower part of the nose is made up of cartilage).

Zygomatic: a pair of bones commonly known as the cheekbones. They also form a large portion of the lateral walls of the eye orbits.

Maxillae: the two maxillary bones fuse to form the upper jaw. The upper teeth are imbedded in the maxillae.

Mandible: the lower jawbone is the strongest bone in the face; it joins the temporal bones on each side of the face, forming the only freely movable joints in the skull. The horizontal part of the mandible, or the *body*, forms the chin. Two upright bars of bone, or *rami*, extend from the body to connect the mandible with the temporal bone. The lower teeth are imbedded in the mandible.

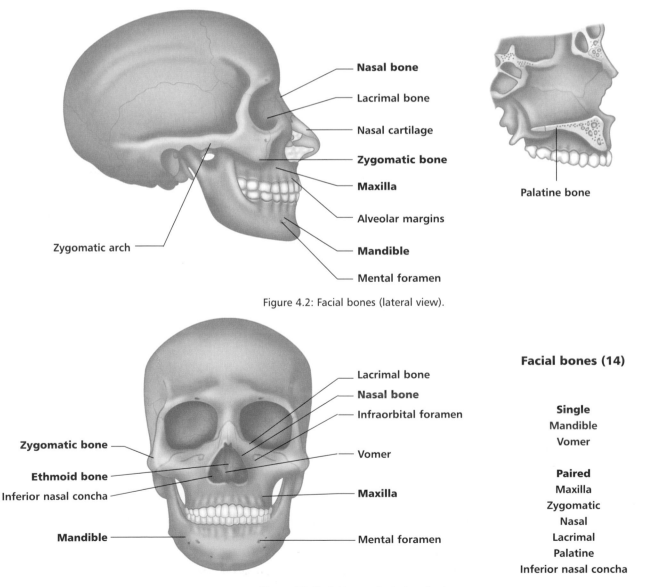

Figure 4.2: Facial bones (lateral view).

Figure 4.3: Facial bones (anterior view).

Facial bones (14)

Single
Mandible
Vomer

Paired
Maxilla
Zygomatic
Nasal
Lacrimal
Palatine
Inferior nasal concha

The Vertebral Column (Spine)

The vertebral column consists of 33 vertebrae in total:

- 7 cervical vertebrae.
- 12 thoracic vertebrae – which also form joints with the 12 ribs.
- 5 lumbar vertebrae – the largest, weight-bearing vertebrae.
- Sacrum (5 fused) – note that the holes, or foramina, in the sacrum correspond to the original gaps between the vertebrae.
- Coccyx (3–4 fused).

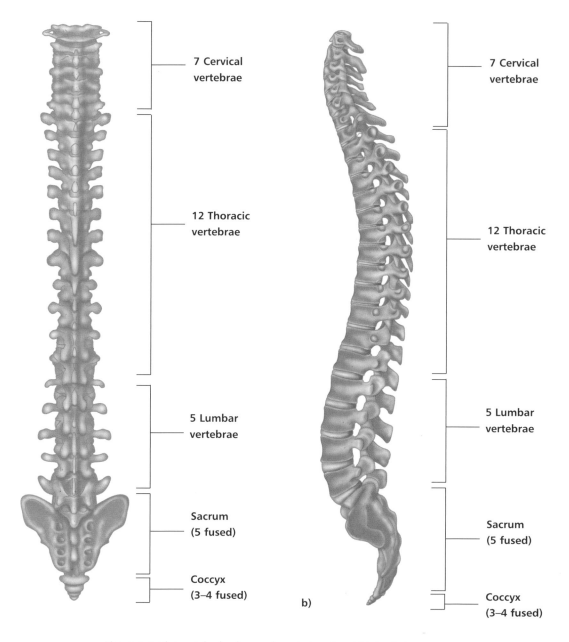

a)

7 Cervical vertebrae

12 Thoracic vertebrae

5 Lumbar vertebrae

Sacrum (5 fused)

Coccyx (3–4 fused)

b)

7 Cervical vertebrae

12 Thoracic vertebrae

5 Lumbar vertebrae

Sacrum (5 fused)

Coccyx (3–4 fused)

Figure 4.4: The vertebral column; a) posterior view, b) lateral view.

A Typical Vertebra

A typical vertebra has the following parts:

Body: the disk-like, weight-bearing part of the vertebra. It faces anteriorly in the vertebral column.

Vertebral arch: the arch formed from the joining of the processes to the body.

Vertebral foramen: the canal through which the spinal cord passes.

Transverse process: two lateral projections.

Spinous process: a single projection that rises from the posterior part of the vertebral arch. On the cervical vertebrae, the spinous processes are short and divide into two points (it looks a little like a whale's tail). On the thoracic vertebrae, the spinous processes are single, slender points that angle sharply downward. On the lumbar vertebrae, the spinous processes are thick and wedge shaped.

Superior and inferior articular processes: paired projections lateral to the foramen. They allow one vertebra to form a joint with the next vertebra.

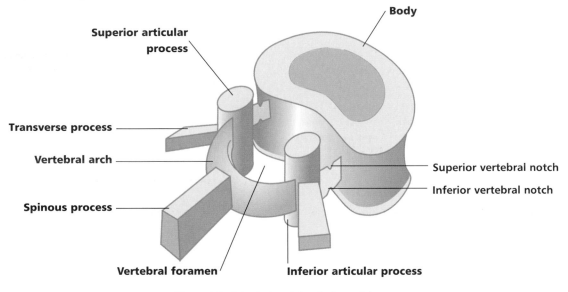

Figure 4.5: A typical vertebra (schematic).

The illustrations on the following pages give a selection of key vertebrae shown from various angles, to depict their variation in shape and features.

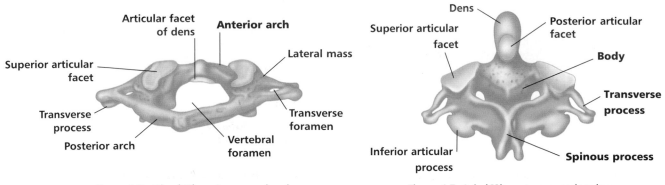

Figure 4.6: Atlas (C1) postero-superior view.

Figure 4.7: Axis (C2) postero-superior view.

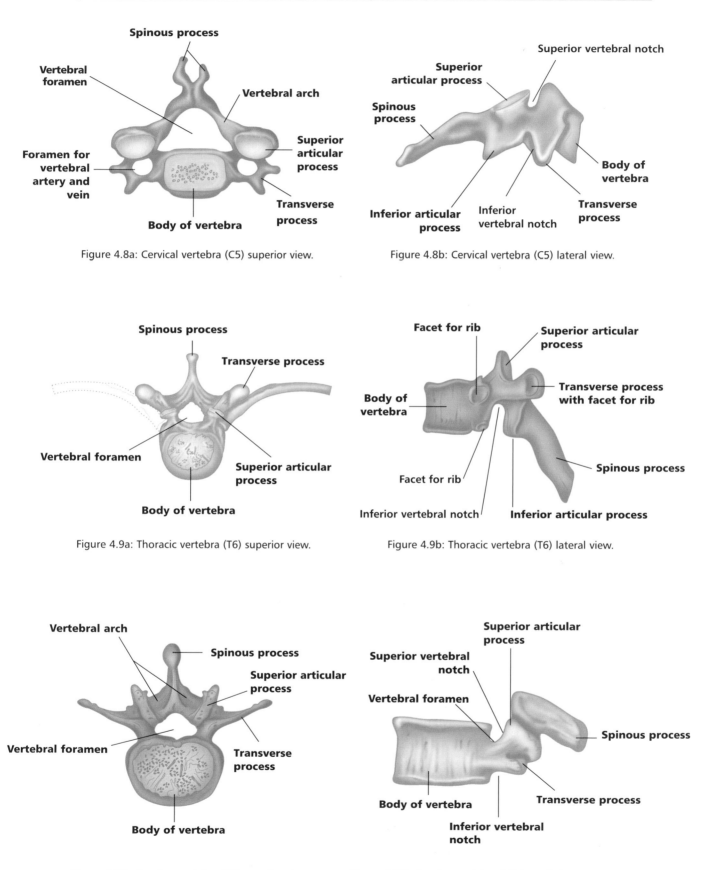

Figure 4.8a: Cervical vertebra (C5) superior view.

Figure 4.8b: Cervical vertebra (C5) lateral view.

Figure 4.9a: Thoracic vertebra (T6) superior view.

Figure 4.9b: Thoracic vertebra (T6) lateral view.

Figure 4.10a: Lumbar vertebra (L3) superior view.

Figure 4.10b: Lumbar vertebra (L3) lateral view.

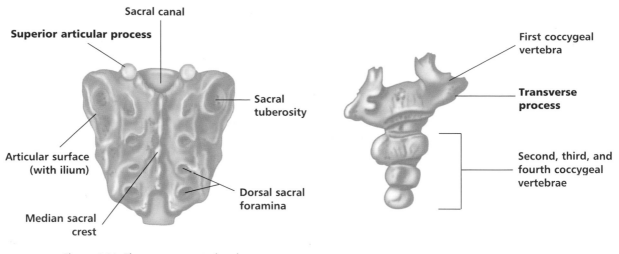

Figure 4.11: The sacrum: posterior view.

Figure 4.12: The coccyx: posterior view.

The Bony Thorax

Sternum

The sternum is commonly known as the breastbone. It is actually the fusion of three bones: the manubrium, the body (also known as the *gladiolus*), and the xyphoid process.

NOTE: The sternum is attached to the first seven pairs of ribs by the costal cartilage. Manubrium means *handle*, as in the handle of a sword; Xiphoid means *sword shaped*.

The Ribs

The ribs consist of 12 pairs in total (comprising true, false and floating ribs).

• True ribs: the first seven pairs attach by costal cartilage directly to the sternum.
• False ribs: the next three pairs attach to costal cartilage but not directly to the sternum.
• Floating ribs: the last two pairs of ribs lack attachment either to costal cartilage or to the sternum.

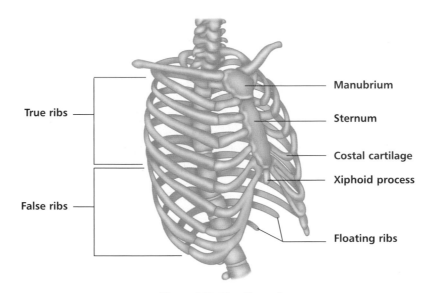

Figure 4.13: The ribs and sternum.

The Appendicular Skeleton

5

The Pectoral Girdle

NOTE: Some soft tissues (e.g. cartilages, aponeuroses, ligaments and tendons) are included where appropriate, for ease of reference.

Clavicle
The clavicle is commonly known as the collarbone; a slender, doubly curved bone that attaches to the manubrium of the sternum medially (the *sternoclavicular joint*) and to the acromion of the scapula laterally (the *acromioclavicular joint*).

Scapula
Commonly known as the shoulder blade; the scapula is a large triangular flat bone lying posterior to the dorsal thorax between the second and seventh ribs. Each scapula articulates with the clavicle and the humerus. The scapula has four important bone markings:

1. The spine – a sharp, prominent ridge on the posterior surface of the scapula that can be easily felt through the skin.
2. The acromion – an enlarged anterior projection at the lateral end of the spine of scapula that can be felt as the 'point of the shoulder'.
3. The corocoid process – projecting forward from the upper border of the scapula.
4. The glenoid fossa – a shallow depression at the lateral angle of the scapula that articulates with the head of the humerus.

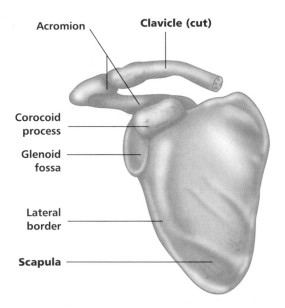

Figure 5.1a: The clavicle and scapula (anterior view).

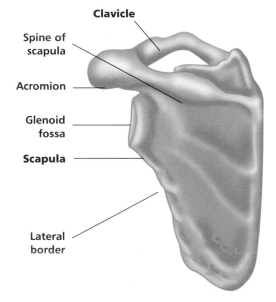

Figure 5.1b: The scapula (posterior view).

The Upper Limb

Humerus

The humerus (arm bone) is the longest and largest bone of the upper limb. It articulates proximally with the scapula (at the glenoid fossa). At the distal end are the *trochlea* (which looks like a spool) and the *capitulum* (or head), which form part of the elbow joint with the ulna and the radius. On either side of the trochlea are the medial and lateral epicondyles of the humerus, easily felt superficially.

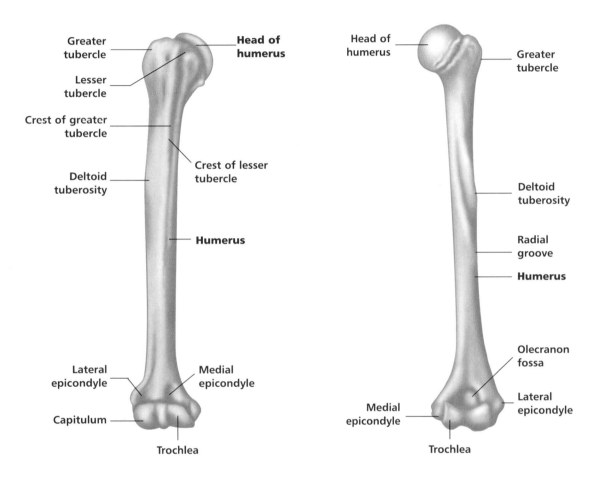

Figure 5.2a: The right humerus (anterior view).   Figure 5.2b: The right humerus (posterior view).

Radius

The radius is one of the two bones in the forearm, on the lateral, or thumb side of the forearm. Proximally the head of the radius forms a joint with the capitulum of the humerus. The radius crosses the ulna during pronation.

Ulna

The ulna is the medial bone in the forearm, on the little finger side. At the proximal end of the ulna are two processes: the *coronoid* and the *olecranon*, which fit over the two medial rounded spools of the trochlea of the humerus. The olecranon is the pointed bump felt when the elbow is bent, and is also known as 'the funny bone', because when the nerve that runs over the olecranon is hit, it can be painful. The *styloid processes* of the radius and ulna can be felt as sharp projections on either side of the wrist.

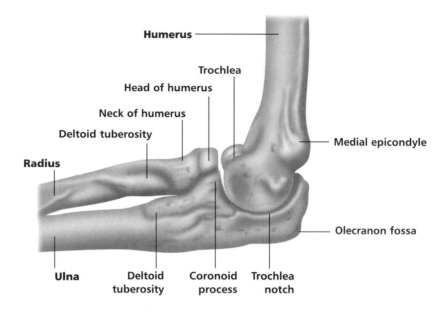

Figure 5.3: Right elbow: medial view in 90 degrees flexion.

8 Carpals

The eight small carpal bones make up the wrist. They are bound together by ligaments and are arranged in two transverse rows, four bones to a row. The first row comprises the scaphoid, lunate, triquetrum, pisiform. The second row comprises the trapezium, trapezoid, capitate, hamate. A mnemonic for memorizing the carpals from lateral to medial, beginning with the proximal row is: "some lovers try positions that they can't handle".

5 Metacarpals

The metacarpals are five bones running between the wrist and the knuckles (which are the heads of the metacarpals).

14 Phalanges

Each finger has three phalanges, whereas the thumb only has two.

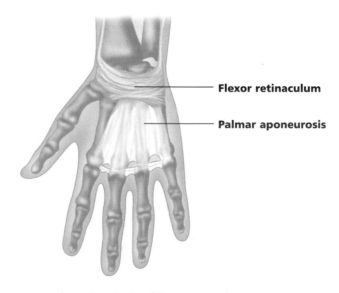

Figure 5.4: The hand (anterior view).

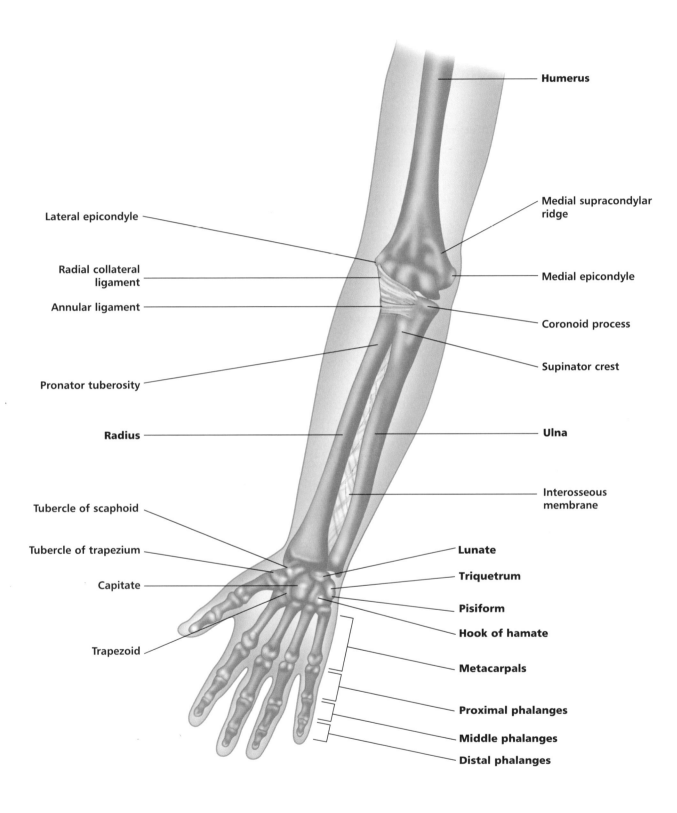

Figure 5.5: The bones of the right forearm and hand (anterior view).

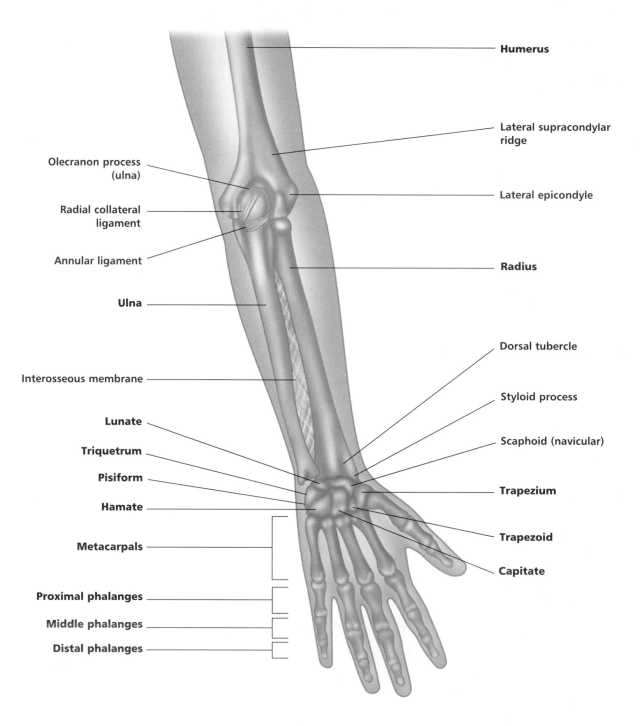

Humerus

Lateral supracondylar ridge

Olecranon process (ulna)

Lateral epicondyle

Radial collateral ligament

Annular ligament

Radius

Ulna

Dorsal tubercle

Interosseous membrane

Styloid process

Lunate

Scaphoid (navicular)

Triquetrum

Pisiform

Trapezium

Hamate

Metacarpals

Trapezoid

Proximal phalanges

Capitate

Middle phalanges

Distal phalanges

Figure 5.6: The bones of the right forearm and hand (posterior view).

The Pelvic Girdle (Os Innominatum)

The pelvic girdle (or hip girdle) consists of two pelvic, or coxal bones. It provides a strong and stable support for the lower extremities on which the weight of the body is carried. The pelvic bones unite with one another in the front (anteriorly) at the *pubic symphysis* (a fibrocartilage disc). With the sacrum and coccyx, the two pelvic bones form a basin-like structure called the pelvis. At birth, each pelvic or coxal bone consists of three separate bones: the ilium, the ischium, and the pubis. These separate bones eventually fuse into one pelvic bone, and the area where they join is a deep hemispherical socket called the *acetabulum* (this socket articulates with the head of the femur). Although the pelvic bone is one bone, it is still commonly discussed as if it consisted of three portions.

Ilium

The ilium is a large, flaring bone that forms the largest and most superior portion of the pelvic bone. *Iliac crests* are felt when you rest your hands on your hips. Each crest terminates in the front as the *Anterior Superior Iliac Spine* or ASIS; and at the back as the *Posterior Superior Iliac Spine* or PSIS (the PSIS is difficult to palpate, but its position is revealed by a skin dimple in the sacral region, level approximately to the second sacral foramen).

Ischium

The ischium is the inferior, posterior part of the pelvic bone, roughly arch-shaped. At the bottom of the ischium are the roughened and thickened *ischial tuberosities* (sometimes called the 'sit-bones', because when we sit, our weight is borne entirely by the ischial tuberosities).

Pubis

The pubis is the anterior and inferior part of the pelvic bone.

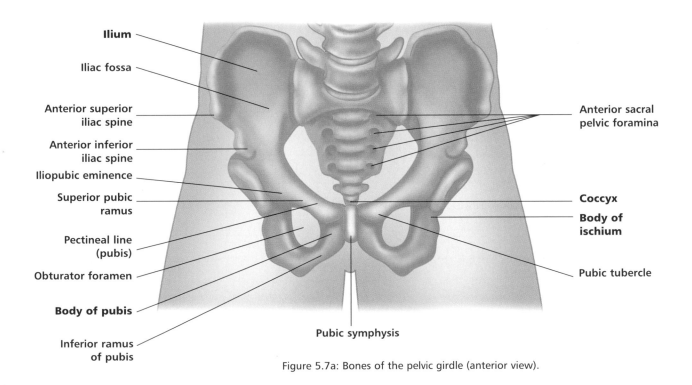

Figure 5.7a: Bones of the pelvic girdle (anterior view).

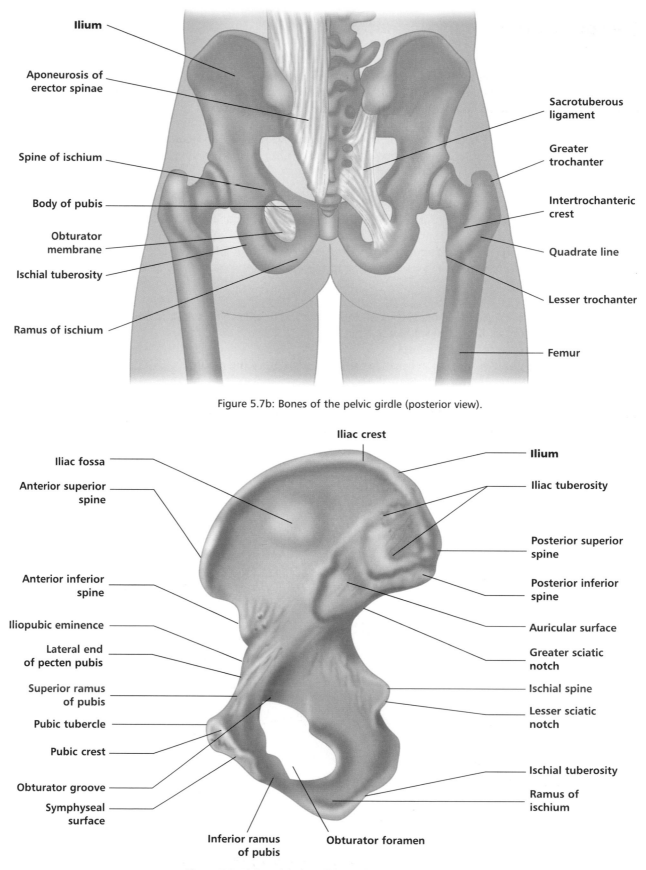

Ilium

Aponeurosis of
erector spinae

Spine of ischium

Body of pubis

Obturator
membrane

Ischial tuberosity

Ramus of ischium

Sacrotuberous
ligament

Greater
trochanter

Intertrochanteric
crest

Quadrate line

Lesser trochanter

Femur

Figure 5.7b: Bones of the pelvic girdle (posterior view).

Iliac crest

Iliac fossa

Anterior superior
spine

Anterior inferior
spine

Iliopubic eminence

Lateral end
of pecten pubis

Superior ramus
of pubis

Pubic tubercle

Pubic crest

Obturator groove

Symphyseal
surface

Inferior ramus
of pubis

Obturator foramen

Ilium

Iliac tuberosity

Posterior superior
spine

Posterior inferior
spine

Auricular surface

Greater sciatic
notch

Ischial spine

Lesser sciatic
notch

Ischial tuberosity

Ramus of
ischium

Figure 5.8: Right pelvic (coxal) bone (posterior view).

The Lower Limb

Femur

The femur is the only bone in the thigh. It is the heaviest, longest and strongest bone in the body. Its proximal end has a ball-like head that articulates with the pelvic bone at the acetabulum. Distally on the femur are the lateral and medial condyles, which articulate with the tibia.

– the *greater trochanter* is a projection just distal to the head and neck of the femur and can sometimes be felt in the buttock.

Tibia

The tibia (shin bone) is the larger and more medial of the bones in the lower leg. At the proximal end, the *medial and lateral condyles* articulate with the distal end of the femur to form the knee joint.

– the *tibial tuberosity* is a roughened area on the anterior surface of the tibia.
– the *medial malleolus* can be felt as the inner bulge of the ankle.

Fibula

The fibula lies lateral and parallel to the tibia and is thin and sticklike. The fibula is not a weight bearing bone. It also plays no part in the knee joint.

– the *lateral malleolus* on the fibula can be felt as the outer bulge of the ankle.

Patella

Known as the 'knee cap', the patella is a small triangular sesamoid bone within the tendon of the quadriceps femoris muscle. It forms the front of the knee joint.

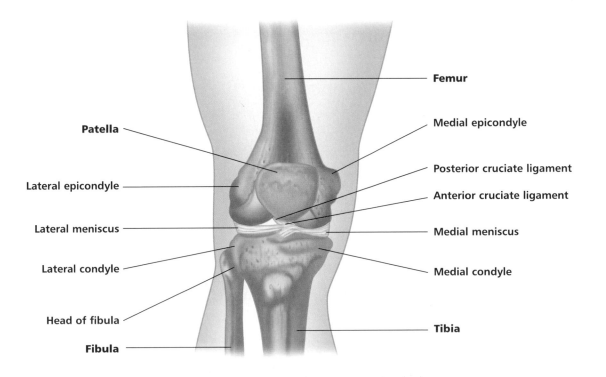

Figure 5.9: Lower femur and upper tibia / fibula of the right leg (anterior view).

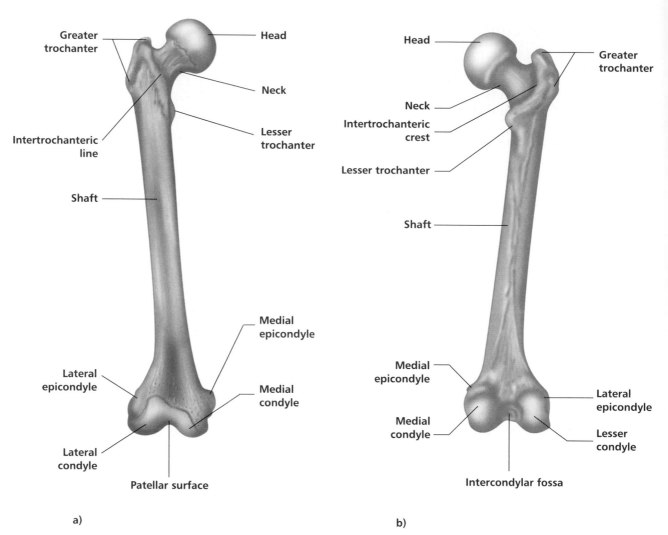

Figure 5.10: Femur of the right leg; a) anterior view, b) posterior view.

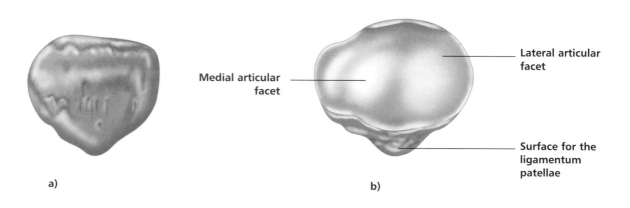

Figure 5.11: Patella of the right leg; a) anterior view, b) posterior view.

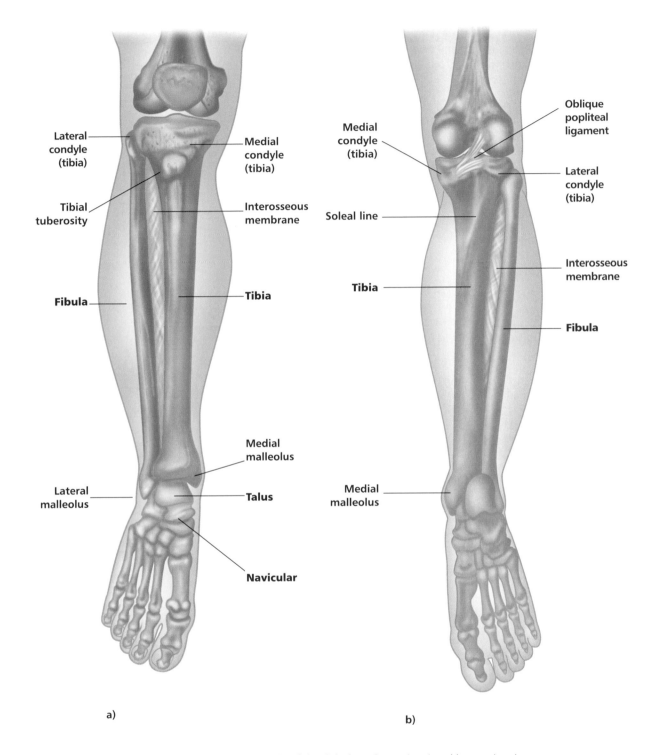

Lateral condyle (tibia)

Medial condyle (tibia)

Tibial tuberosity

Interosseous membrane

Fibula

Tibia

Lateral malleolus

Medial malleolus

Talus

Navicular

Medial condyle (tibia)

Oblique popliteal ligament

Lateral condyle (tibia)

Soleal line

Interosseous membrane

Tibia

Fibula

Medial malleolus

a)

b)

Figure 5.12: Tibia and fibula of the right leg, a) anterior view, b) posterior view.

7 Tarsals

The tarsals are the seven bones of the ankle. The two largest tarsals mostly carry the body weight: the *calcaneus*, or the heel bone, and the *talus*, which lies between the tibia and the calcaneus. The navicular, medial cuneiform, intermediate cuneiform, lateral cuneiform and cuboid constitute the other five tarsals.

5 Metatarsals

The metatarsals form the instep or sole of the foot.

14 Phalanges

Each toe has three phalanges, except the big toe, which has only two.

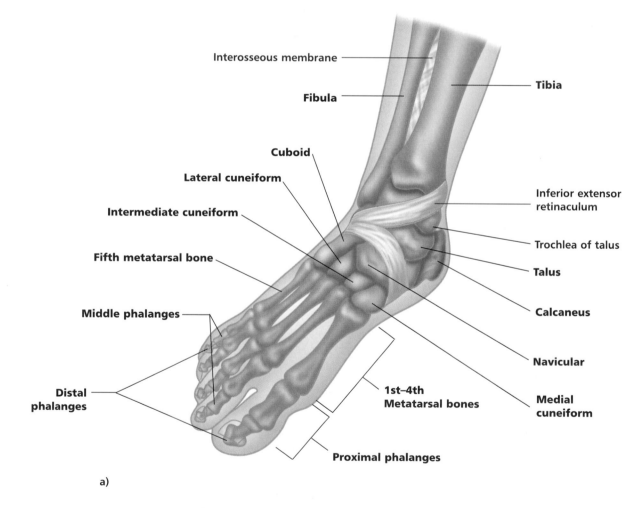

Figure 5.13: Bones of the right foot; a) anteromedial view.

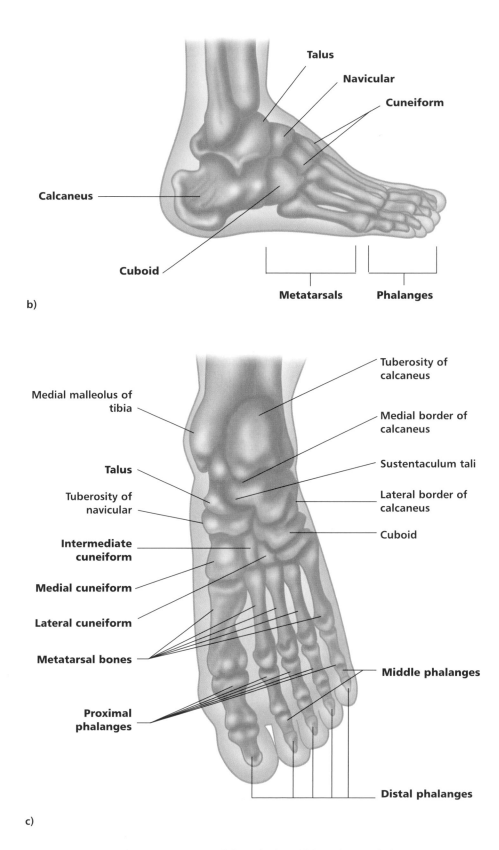

b)

c)

Figure 5.13: Bones of the right foot; b) lateral view, c) plantar view.

General Skeletal Interrelationships

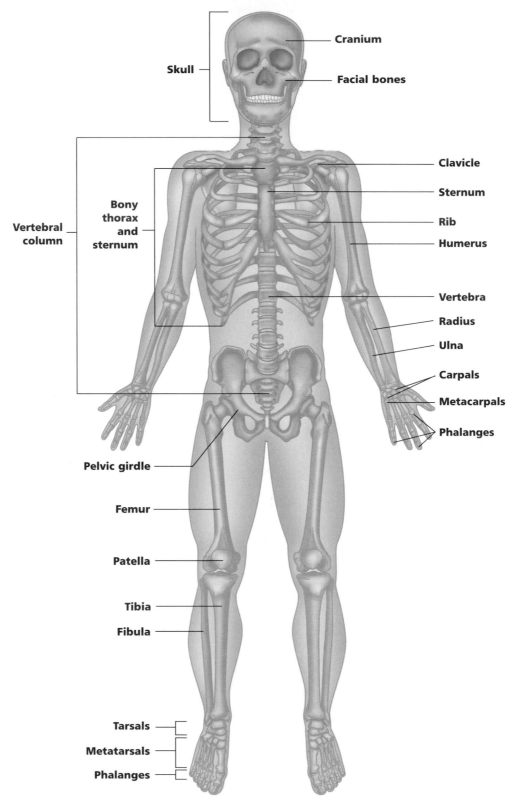

Figure 5.14: Skeleton (anterior view).

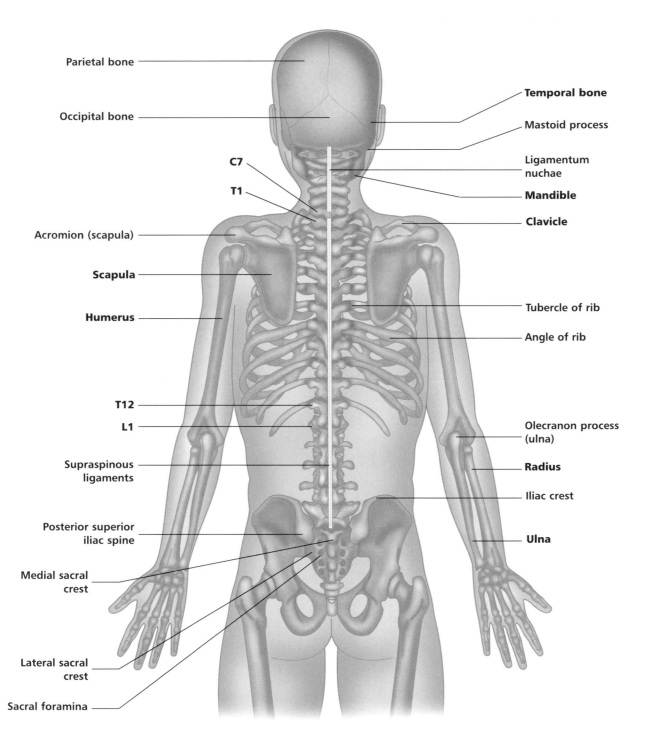

Parietal bone

Occipital bone

Temporal bone

Mastoid process

Ligamentum nuchae

Mandible

C7

T1

Clavicle

Acromion (scapula)

Scapula

Humerus

Tubercle of rib

Angle of rib

T12

L1

Olecranon process (ulna)

Radius

Supraspinous ligaments

Iliac crest

Posterior superior iliac spine

Ulna

Medial sacral crest

Lateral sacral crest

Sacral foramina

Figure 5.15: Skeleton (posterior view).

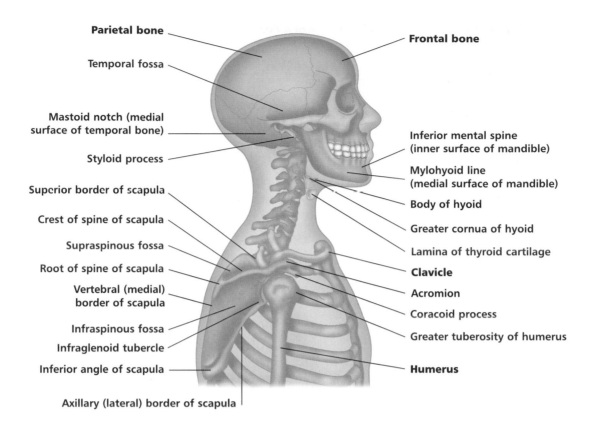

Parietal bone

Temporal fossa

Mastoid notch (medial surface of temporal bone)

Styloid process

Superior border of scapula

Crest of spine of scapula

Supraspinous fossa

Root of spine of scapula

Vertebral (medial) border of scapula

Infraspinous fossa

Infraglenoid tubercle

Inferior angle of scapula

Axillary (lateral) border of scapula

Frontal bone

Inferior mental spine (inner surface of mandible)

Mylohyoid line (medial surface of mandible)

Body of hyoid

Greater cornua of hyoid

Lamina of thyroid cartilage

Clavicle

Acromion

Coracoid process

Greater tuberosity of humerus

Humerus

Figure 5.16: Skull to humerus (lateral view).

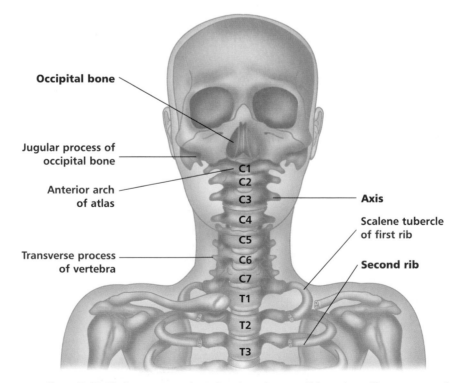

Occipital bone

Jugular process of occipital bone

Anterior arch of atlas

Transverse process of vertebra

C1
C2
C3
C4
C5
C6
C7
T1
T2
T3

Axis

Scalene tubercle of first rib

Second rib

Figure 5.17: Skull to sternum (anterior view, the mandible and maxilla are removed).

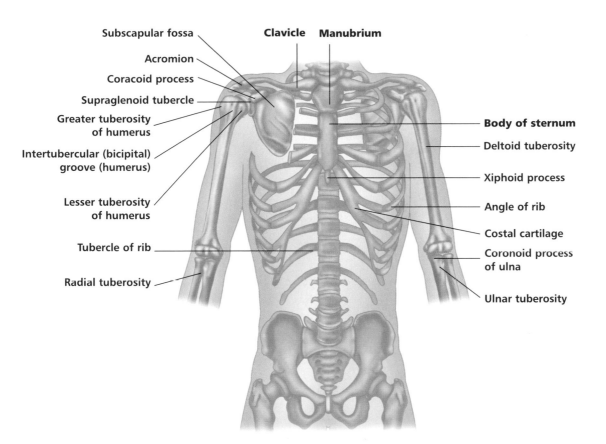

Subscapular fossa
Acromion
Coracoid process
Supraglenoid tubercle
Greater tuberosity
of humerus
Intertubercular (bicipital)
groove (humerus)
Lesser tuberosity
of humerus
Tubercle of rib
Radial tuberosity

Clavicle   **Manubrium**

Body of sternum
Deltoid tuberosity
Xiphoid process
Angle of rib
Costal cartilage
Coronoid process
of ulna
Ulnar tuberosity

Figure 5.18: Ribcage, pectoral girdle, upper arm (anterior view, the upper right anterior ribcage is removed).

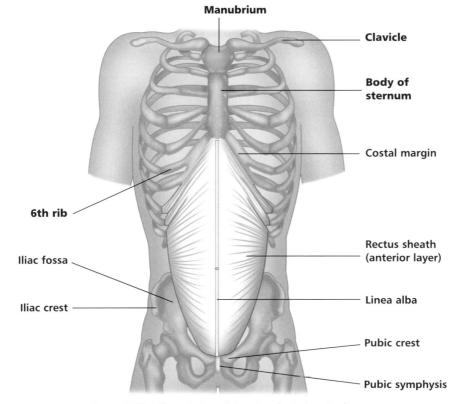

Manubrium

Clavicle

**Body of
sternum**

Costal margin

6th rib

Iliac fossa

Iliac crest

Rectus sheath
(anterior layer)

Linea alba

Pubic crest

Pubic symphysis

Figure 5.19a: Thoracic to pelvic region (anterior view).

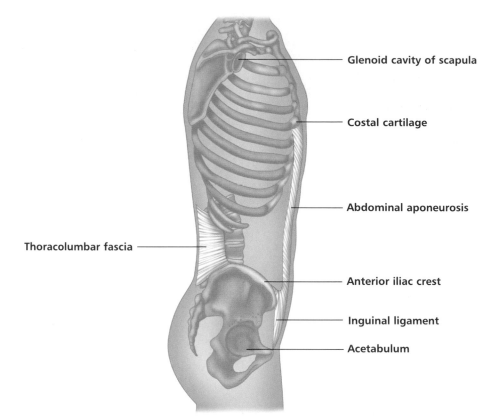

Glenoid cavity of scapula

Costal cartilage

Abdominal aponeurosis

Thoracolumbar fascia

Anterior iliac crest

Inguinal ligament

Acetabulum

Figure 5.19b: Thoracic to pelvic region (lateral view).

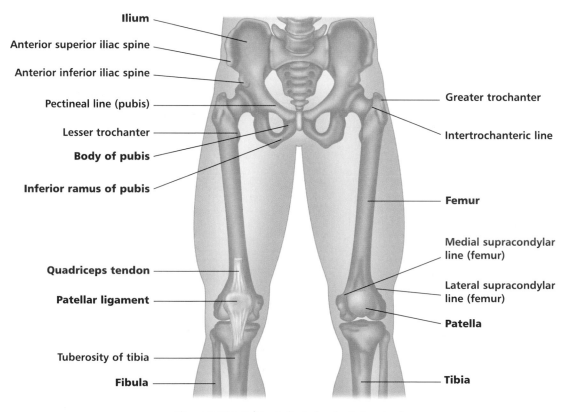

Ilium

Anterior superior iliac spine

Anterior inferior iliac spine

Pectineal line (pubis)

Lesser trochanter

Body of pubis

Inferior ramus of pubis

Greater trochanter

Intertrochanteric line

Femur

Medial supracondylar line (femur)

Lateral supracondylar line (femur)

Patella

Quadriceps tendon

Patellar ligament

Tuberosity of tibia

Fibula

Tibia

Figure 5.20a: Pelvic girdle to leg (anterior view).

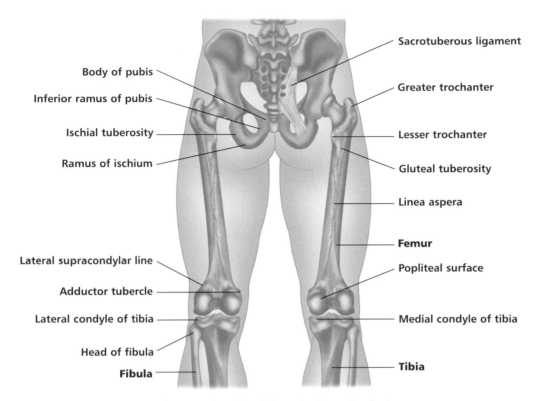

Body of pubis

Inferior ramus of pubis

Ischial tuberosity

Ramus of ischium

Lateral supracondylar line

Adductor tubercle

Lateral condyle of tibia

Head of fibula

Fibula

Sacrotuberous ligament

Greater trochanter

Lesser trochanter

Gluteal tuberosity

Linea aspera

Femur

Popliteal surface

Medial condyle of tibia

Tibia

Figure 5.20b: Pelvic girdle to leg (posterior view).

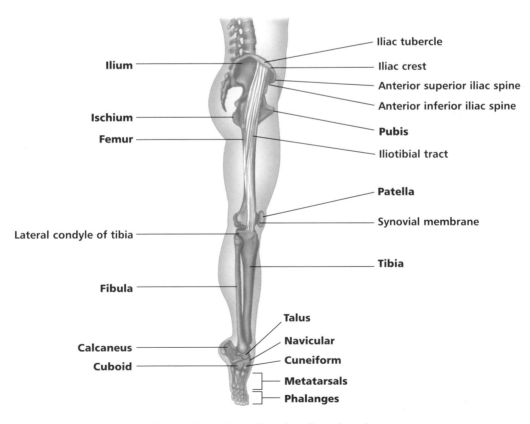

Ilium

Ischium

Femur

Lateral condyle of tibia

Fibula

Calcaneus

Cuboid

Iliac tubercle

Iliac crest

Anterior superior iliac spine

Anterior inferior iliac spine

Pubis

Iliotibial tract

Patella

Synovial membrane

Tibia

Talus

Navicular

Cuneiform

Metatarsals

Phalanges

Figure 5.21: Pelvic girdle to foot (lateral view).

Bony Landmarks Seen or Felt Near the Body Surface

The following bony landmarks can be seen or felt near the surface of the body. Identify them on yourself or a partner, using figure 5.22 (a–c) for reference.

Frontal bone
Temporal bone
Occipital bone
Manubriosternum and
  manubriosternal joint
  (level with the 2nd rib)
2nd rib
Sternoclavicular joint
Acromioclavicular joint
Spine of the scapula
Medial border of scapula
Inferior angle of scapula
Medial and lateral epicondyle
  of humerus
Olecranon
Head of the radius
Ulnar styloid
Pisiform bone
Anatomical snuffbox
Iliac crest
Anterior superior iliac spine (ASIS)
Posterior superior iliac spine (PSIS)
Ischial tuberosities
Greater trochanter
Head of the fibula
Tibial tuberosity
Medial and lateral malleolus
Calcaneus
Spinous process of the vertebrae

Hints

C2: the first cervical vertebra to be felt below the occiput.
C7: at the base of the neck, the vertebra that stands out most prominently.
T3–4: level with the spine of the scapula.
T7: level with the inferior angle of the scapula.
L4: level with the iliac crest.
S2: level with the PSIS (or visible as the dimple at the top of the buttocks).

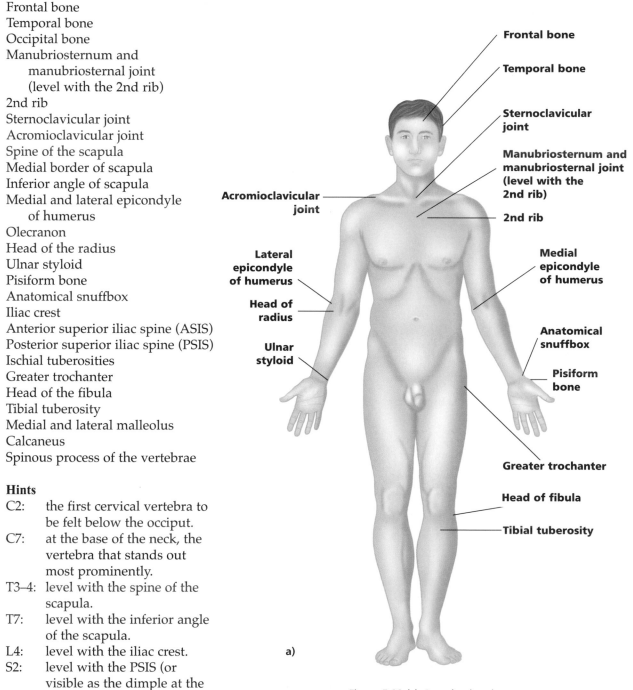

a)

Figure 5.22 (a): Bony landmarks.

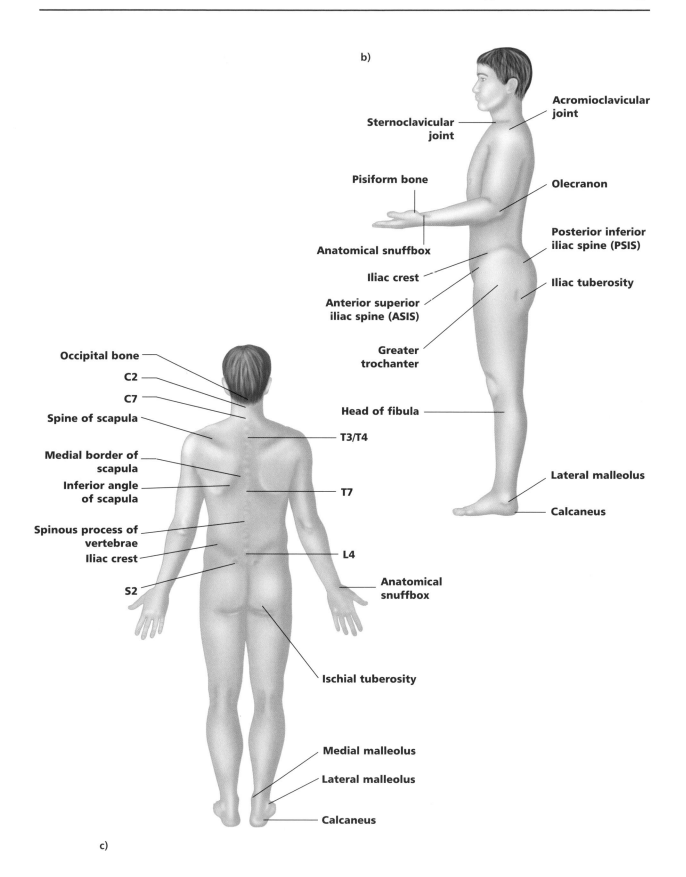

Figure 5.22 (b–c): Bony landmarks.

Joints

6

With the exception of the hyoid bone in the neck, all other bones form a joint with at least one other bone. Joints are also called *articulations*.

Joints have two functions: to *hold the bones together*, and to *give the rigid skeleton mobility*. When two bones meet, or articulate, there may or may not be movement depending on, (a) the amount of bonding material between the bones; (b) the nature of the material between the bones; (c) the shape of the bony surfaces; (d) the amount of tension in the ligaments or muscles involved in the joint; (e) the position of the ligaments and muscles.

PART ONE – Classification of Joints

Joints are classified in two ways: **functionally** and **structurally**.

Functionally

The functional classification of joints focuses on the amount of movement allowed by the joint.

Immovable Joints (Synarthrotic)

These joints are found mostly in the axial skeleton, where joint stability and firmness is important for the protection of the internal organs.

Slightly Movable Joints (Amphiarthrotic)

Like immovable joints, and for the same reason, these joints are also found mainly in the axial skeleton.

Freely Movable Joints (Diarthrotic)

These joints predominate in the limbs, where a greater range of movement is required.

Structurally

Fibrous Joints

In fibrous joints, fibrous tissue joins the bones. As such, no joint cavity is present. Generally these joints have little or no movement, i.e. they are *synarthrotic*. Fibrous joints are of three types: *sutures*, *syndesmoses*, and *gomphoses*.

1. Sutures

The only examples of fibrous sutures are the sutures of the skull, where the irregular edges of the bones interlock and are bound tightly together by connective tissue fibres, allowing no active movement. Layers of periosteum on the inner and outer layers of the adjoining bones bridge the gap between the bones and form the main bonding factor. Between the adjoining joint surfaces there is a layer of vascular fibrous tissue that also helps unite the bones. This vascular fibrous tissue, along with the two layers of periosteum, is collectively called the *sutural ligament*. The fibrous tissue becomes ossified during adulthood by a process that occurs first at the deep aspect of the suture, progressively extending to the superficial part. This ossifying process is referred to as *synostosis*.

2. Syndesmoses

A syndesmosis is a fibrous joint where the uniting fibrous tissue forms an *interosseous membrane* or *ligament*; i.e. a band of fibrous tissue that allows little movement, situated between the radius and ulna and between the tibia and fibula.

3. Gomphoses

A gomphosis refers to a fibrous joint in which a peg is embedded into a socket. The only examples of such joints in humans consist of the teeth fixed into their sockets.

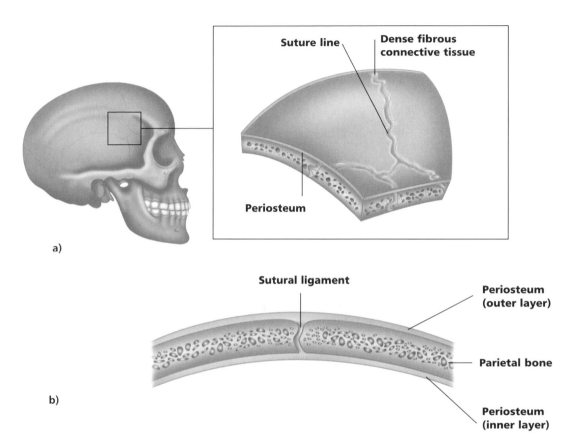

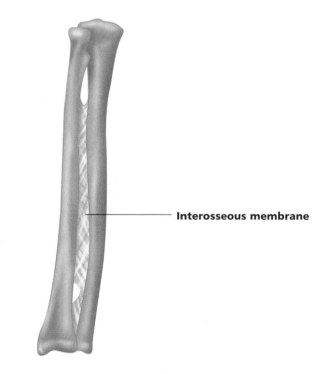

Figure 6.1: a) Position of a suture; b) vertical section through a suture.

Figure 6.2: The interosseous membrane between the radius and ulna.

Cartilaginous Joints

In cartilaginous joints, a continuous plate of hyaline cartilage or a fibrocartilage disc connects the bones. Again, no joint cavity is present. They can be either immovable (synchondrosis) or slightly movable (symphysis). The slightly movable joints are the more common.

Synchondroses

Examples of cartilaginous joints that are immovable are the epiphyseal plates of growing long bones. These plates are made of hyaline cartilage that ossifies in young adults (*see* page 29). Thus, the place where a joint is united by such a plate is known as a *synchondrosis*. Another example of such a joint that eventually ossifies is the joint between the first rib and the manubrium of the sternum.

Symphyses

Examples of slightly movable cartilaginous joints are the pubic symphysis of the pelvic girdle, and the intervertebral joints of the spinal column. In both cases the articular surfaces of the bones are covered with hyaline cartilage that is in turn fused to a 'pad' of fibrocartilage (fibrocartilage is compressible and resilient, and acts as a shock absorber).

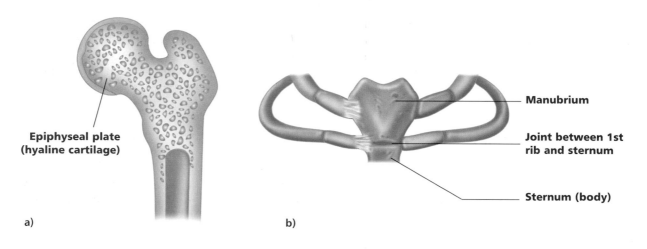

**Epiphyseal plate
(hyaline cartilage)**

Manubrium

**Joint between 1st
rib and sternum**

Sternum (body)

a)   b)

Figure 6.3: Cartilaginous immovable (synchondroses) joints (anterior view); a) the epiphyseal plate in a growing long bone, b) the sternocostal joint between manubrium and first rib.

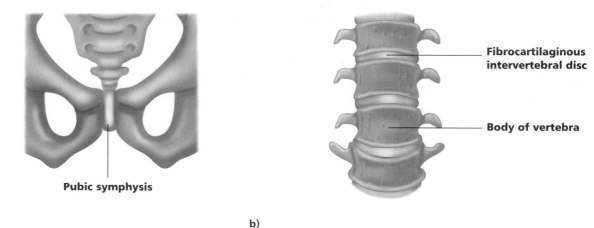

**Fibrocartilaginous
intervertebral disc**

Body of vertebra

Pubic symphysis

a)   b)

Figure 6.4: Cartilaginous slightly movable (amphiarthrotic / symphysis) joints (anterior view);
a) pubic symphysis of the pelvic girdle, b) intervertebral joints.

Synovial Joints

Synovial joints possess a joint cavity that contains *synovial fluid*. They are freely movable, *diarthrotic*, joints. Synovial joints have a number of distinguishing features:

Articular cartilage (or *hyaline cartilage*): covers the ends of the bones that form the joint.

A joint cavity: this cavity is more a potential space than a real one, because it is filled with lubricating *synovial fluid*. The joint cavity is enclosed by a double-layered 'sleeve' or capsule known as the *articular capsule*.

The external layer of the articular capsule is known as the *capsular ligament*. It is a tough, flexible, fibrous connective tissue that is continuous with the periostea of the articulating bones. The internal layer, or *synovial membrane*, is a smooth membrane made of loose connective tissue that lines the capsule and all internal joint surfaces other than those covered in hyaline cartilage.

Synovial fluid: a slippery fluid that occupies the free spaces within the joint capsule. Synovial fluid is also found within the articular cartilage and provides a film that reduces friction between the cartilages. When a joint is compressed by movement the fluid is forced out of the cartilage; when pressure is relieved the fluid rushes back into the articular cartilage. Synovial fluid nourishes the cartilage, which is *avascular* (contains no blood vessels); it also contains *phagocytic cells* (cells that eat dead matter) that rid the joint cavity of microbes or cellular waste. The amount of synovial fluid varies in different joints, but is always sufficient to form a thin film to reduce friction. During injury to the joint extra fluid is produced and creates the characteristic swelling of the joint. The synovial membrane later reabsorbs this extra fluid.

Collateral or accessory ligaments: synovial joints are reinforced and strengthened by a number of ligaments. These ligaments are either *capsular*, i.e. thickened parts of the fibrous capsule itself, or independent *collateral* ligaments that are distinct from the capsule. Ligaments always bind *bone to bone* and according to their position and quantity around the joint, they will restrict movement in certain directions, and prevent unwanted movement. As a general rule, the more ligaments a joint has, the stronger it is.

Bursae (sing. *bursa*) are fluid-filled sacs often found cushioning the joint. They are lined by synovial membrane and contain synovial fluid. They are found between tendons and bone, ligament and bone, or muscle and bone, and reduce friction by acting as a cushion.

Tendon sheaths are also frequently found in close proximity to synovial joints. They have the same structure as a bursa, and wrap themselves around tendons subject to friction, in order to protect them.

Articular discs (menisci) are present in some synovial joints. They act as shock absorbers (similar to the fibrocartilagenous disc in the pubic symphysis). For example, in the knee joint, two crescent-shaped fibrocartilage discs called the *medial* and *lateral menisci* lie between the medial and lateral condyles of the femur and the medial and lateral condyles of the tibia.

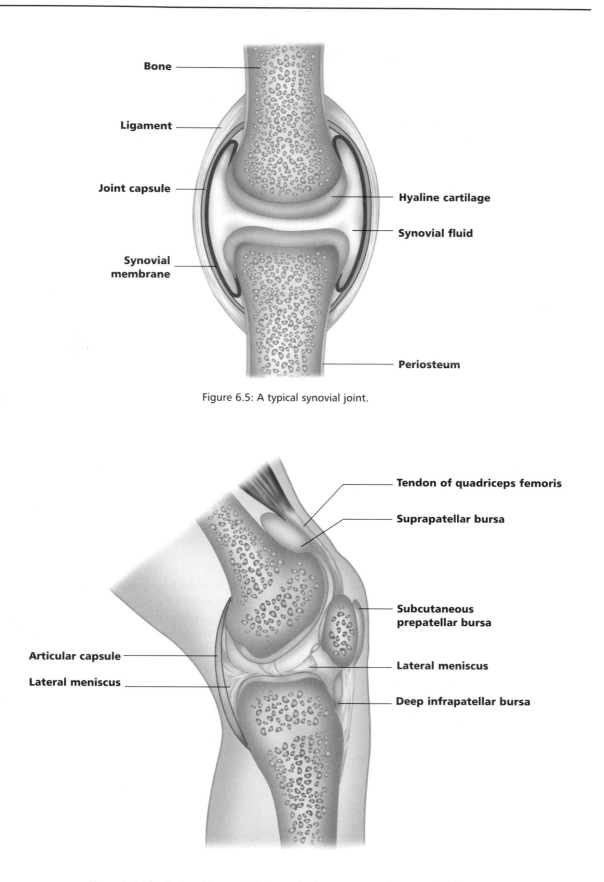

Bone

Ligament

Joint capsule

Hyaline cartilage

Synovial fluid

Synovial membrane

Periosteum

Figure 6.5: A typical synovial joint.

Tendon of quadriceps femoris

Suprapatellar bursa

Subcutaneous prepatellar bursa

Articular capsule

Lateral meniscus

Lateral meniscus

Deep infrapatellar bursa

Figure 6.6: Shock absorbing and friction-reducing structures of a synovial joint.

The Seven Types of Synovial Joints

Plane or Gliding
In gliding joints, movement occurs when two, generally flat or slightly curved, surfaces glide across one another. Examples: the acromioclavicular joint; the joints between the carpal bones in the wrist, or the tarsal bones in the ankle; the facet joints between the vertebrae; the sacroiliac joint.

Hinge
In hinge joints, movement occurs around only one axis; a transverse one – as in the hinge of the lid of a box. A protrusion of one bone fits into a concave or cylindrical articular surface of another, permitting flexion and extension. Examples: the interphalangeal joints, the elbow, and the knee.

Pivot
In pivot joints, movement takes place around a vertical axis, like the hinge of a gate. A more or less cylindrical articular surface of bone protrudes into and rotates within a ring formed by bone or ligament. Examples: the dens of the axis protrude through the hole in the atlas, allowing the rotation of the head from side to side. Also, the joint between the radius and the ulna at the elbow allows the round head of the radius to rotate within a 'ring' of ligament that is secured to the ulna.

Ball and Socket
Ball and socket joints consist of a 'ball' formed by the spherical or hemispherical head of one bone that rotates within the concave socket of another, allowing flexion, extension, adduction, abduction, circumduction, and rotation. Thus, they are multiaxial and allow the greatest range of movement of all joints. Examples: the shoulder and the hip joints.

Condyloid
In common with ball and socket joints, condyloid joints have a spherical articular surface that fits into a matching concavity. Also, like ball and socket joints, condyloid joints permit flexion, extension, abduction, adduction, and circumduction. However, the disposition of surrounding ligaments and muscles prevent active rotation around a vertical axis. Examples: the metacarpophalangeal joints of the fingers (but not the thumb).

Saddle
Saddle joints are similar to condyloid joints, except that both articulating surfaces have convex and concave areas, and so resemble two 'saddles' that join them together by accommodating each other's convex to concave surfaces. Saddle joints allow even more movement than condyloid joints, for example, allowing the 'opposition' of the thumb to the fingers. Example: the carpometacarpal joint of the thumb.

Ellipsoid
An ellipsoid joint is effectively similar to a ball and socket joint, but the articular surfaces are ellipsoid instead of spherical. Movements as for ball and socket joints, with the exception of rotation. The shape of the ellipsoid surfaces prevents this. Example: the radio-carpal joint.

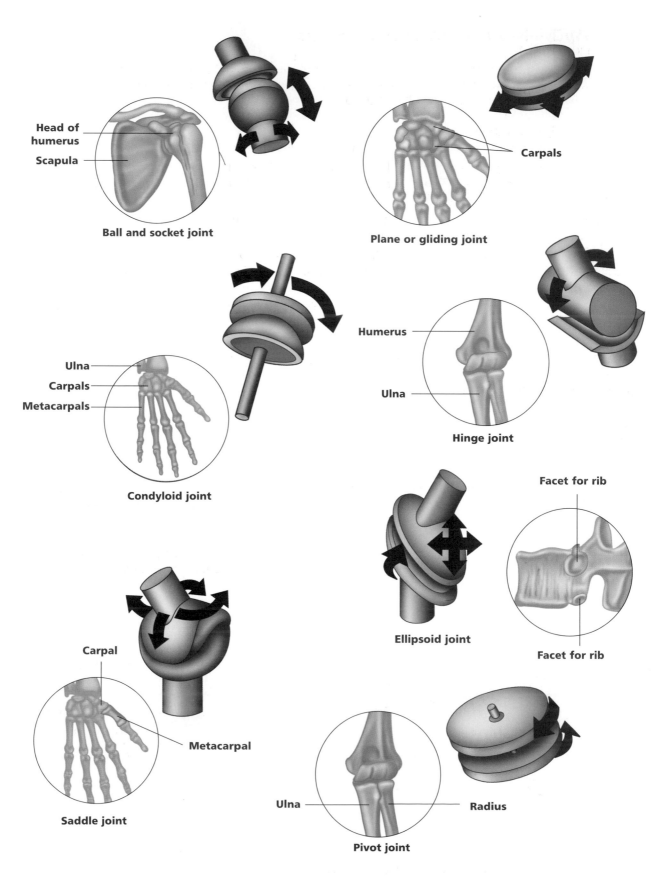

Figure 6.7: Types of synovial joints.

PART TWO – Features of Specific Joints

This section presents a relatively in-depth examination of four synovial joints: i.e. the shoulder, elbow, hip and knee joints. The other joints, of all classifications, are presented in less detail, with minimal accompanying text, but with fully labelled illustrations.

Notes About Synovial Joints:
• Some tendons run partly within the joint and are therefore intracapsular.

• The fibres of many ligaments are largely integrated with those of the capsule and the delineation between capsule and ligament is sometimes unclear. Therefore, only the main ligaments are mentioned.

• Ligaments are termed intracapsular (or intra-articular) when inside the joint cavity, and extracapsular (or extra-articular) when outside the capsule.

• Many ligaments of the knee joint are modified extensions or expansions of muscle tendons, but are classed as ligaments to differentiate them from the more regular stabilizing tendons, such as the patellar ligament from the quadriceps.

• Most synovial joints have various bursae in their vicinity, as shown in the illustrations pertaining to each joint.

Joints of the Head and Vertebral Column

Temporomandibular Joint

Type of Joint
Synovial hinge joint, plus a plane joint.

Articulation
The head of the mandible articulates with the mandibular fossa and the articular tubercle of the temporal bone. A fibrous disc separates the articular surfaces and moulds itself upon them when the joint moves.

Movements
This is the only movable joint in the head. Movement can occur in all three planes: upwards and downwards, backwards and forwards, and from side to side. A gliding action occurs superior to the disc. A hinge action occurs inferior to the disc.

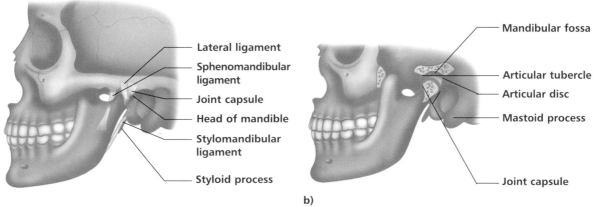

Figure 6.8 (a & b): The temporomandibular joint (lateral view).

Atlanto-occipital Joint

Type of Joint
The articulations of the two sides act together functionally as a synovial ellipsoid joint.

Articulation
Between the occipital condyles and the superior articular facets of the atlas.

Movements
Flexion and extension (as in nodding the head). Lateral flexion.

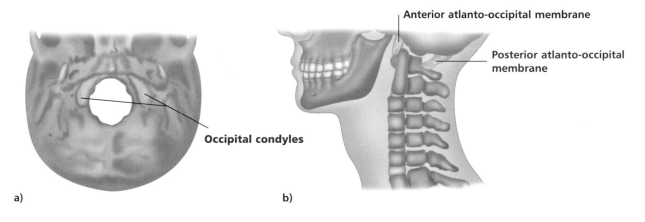

a)   b)

Figure 6.9: The atlanto-occipital joint; a) inferior view, b) lateral view.

Atlanto-axial Joint

Type of Joint
Lateral atlanto-axial joint: Synovial plane.
Median (sagittal) atlanto-axial joint: Synovial pivot.

Articulation
Lateral atlanto-axial joint: Between the opposed articular processes of the atlas and axis.
Median (sagittal) atlanto-axial joint: Between the dens of the axis and the anterior arch of the atlas, and with the transverse ligament.

Movements
Rotation of the head around a vertical axis (the skull and the atlas moving as one).

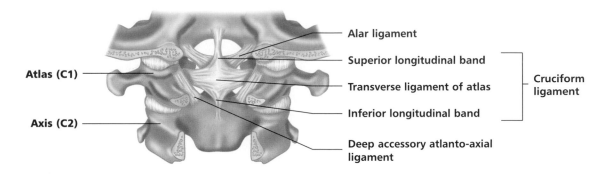

Figure 6.10: The atlanto-axial joint (posterior view).

Joints Between Vertebral Bodies

Type of Joint
Cartilaginous symphysis (slightly movable).

Articulation
Between adjacent surfaces of vertebral bodies, and united by a fibrocartilaginous intervertebral disc.

Movements
Only slight movement occurs between any two successive vertebrae, but there is considerable movement throughout the column as a whole.

Cervical region: Flexion, extension, lateral flexion with rotation, (i.e. lateral flexion cannot occur without an element of rotation and vice versa).
Thoracic region: Rotation, always associated with an element of lateral flexion, and vice versa. Only extremely slight flexion and extension can occur (limited by presence of ribs and sternum).
Lumbar region: Flexion, extension. Only extremely slight rotation can occur (restricted by the angle of the articular processes).

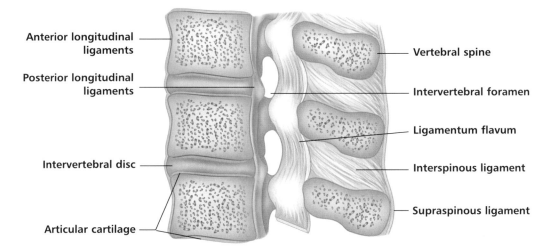

Figure 6.11a: Sagittal section through 2nd to 4th lumbar vertebrae.

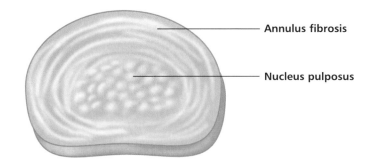

Figure 6.11b: Transverse section of a lumbar intervertebral disc.

Joints Between Vertebral Arches

Type of Joint
Synovial plane.

Articulation
Between opposed articular processes of adjacent vertebrae, to unite adjacent vertebral arches.

Movements
Only slight movement occurs between any two successive vertebrae, but there is considerable movement throughout the column as a whole.

Cervical region: Flexion, extension, lateral flexion with rotation, (i.e. lateral flexion cannot occur without an element of rotation and vice versa).
Thoracic region: Rotation, always associated with an element of lateral flexion, and vice versa. Only extremely slight flexion and extension can occur (limited by presence of ribs and sternum).
Lumbar region: Flexion, extension. Only extremely slight rotation can occur (restricted by the angle of the articular processes).

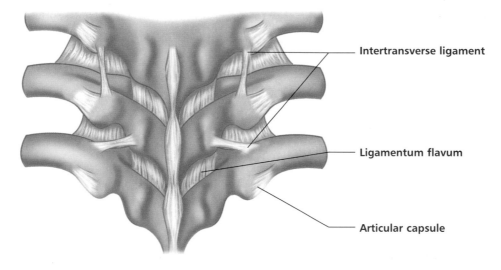

Intertransverse ligament

Ligamentum flavum

Articular capsule

Figure 6.12: A typical vertebral arch joint (posterior view).

Joints of the Ribs and Sternum

Costovertebral Joints

Type of Joint
Joints of the heads of ribs (capitular joints): Synovial plane.
Costotransverse joints: Synovial plane.

Articulation
Joints of the heads of ribs (capitular joints): Superior and inferior articular facets on the head of each typical rib articulate with the facets on two adjacent vertebral bodies (i.e. the rib's head sits between two vertebral bodies, and also against a shallow depression on the intervertebral disc).
Costotransverse joints: The tubercle of each typical rib articulates with the transverse process of the lower of the two vertebrae to which its head is joined (but ligaments attach it to the transverse processes of both vertebrae).

NOTE: The first rib and last two or three ribs have atypical vertebral connections, because the head of these ribs have only one facet, not two; and therefore articulate with one vertebral body rather than two. The tubercles of the lowest ribs do not form synovial joints with the transverse processes.

Movements
The capitular and costotransverse joints of each rib together form a hinge, causing the anterior part of the rib to be raised (with some lateral 'expansion') during inspiration, and lowered (with some medial 'contraction') during expiration. This effectively increases and decreases the anteroposterior and transverse diameters of the thorax with each in-breath and out-breath.

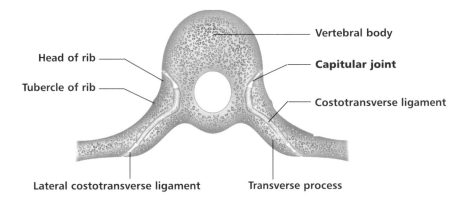

Figure 6.13a: Transverse section through a typical costovertebral joint.

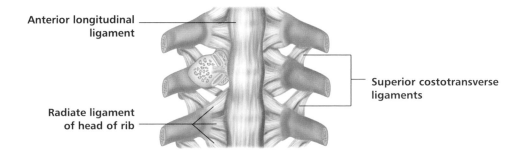

Figure 6.13b: The costovertebral joint (anterior view).

Sternocostal Joints

The hyaline cartilage that is continuous with the anterior end of each rib is called the *costal cartilage*.

Type of Joint

First rib: Cartilaginous immovable (synchondrosis).
Ribs 2–7: Simple synovial plane.
Ribs 8–10: Simple synovial plane articulations at interchondral joints.

Articulation

First rib: Via costal cartilage to the body of the sternum.
Ribs 2–7: Via costal cartilages to facets on the side of the body of the sternum. The joint cavities are divided in two by an intra-articular ligament (until cavities disappear in old age).
Ribs 8–10: Their costal cartilages unite with the costal cartilage of rib 7.
Ribs 11–12: Do not articulate anteriorly, but end freely in the muscles of the flank. They are therefore called floating ribs.

Movements

Enables expansion and contraction of the ribcage, as described under costovertebral joints (*see* page 83).

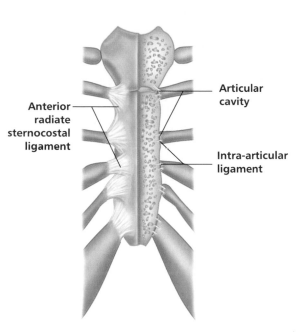

Figure 6.14: The sternocostal joint (anterior view).

Sternal Joints

Type of Joint

Manubriosternal joint: Similar in appearance to a cartilaginous symphysis (slightly movable) joint.
Xiphisternal joint: Cartilaginous immovable (synchondrosis). Usually becomes ossified in old age.

Articulation

Manubriosternal joint: Between the manubrium and body of the sternum, adjacent to the second costal cartilage.
Xiphisternal joint: Between the body of the sternum and the xiphoid process. This joint marks the inferior extent of the thoracic cavity.

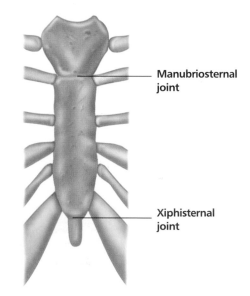

Figure 6.15: The sternal joints (anterior view).

Joints of the Shoulder Girdle and Upper Limb

Sternoclavicular Joint

Type of Joint
Functionally, a synovial ball and socket joint. But unlike most articular surfaces, the articular cartilage is fibrocartilage rather than hyaline cartilage.

Articulation
Between the sternal (medial) end of the clavicle, the clavicular notch of the manubrium, and the costal cartilage of the first rib.

NOTE: A fibrocartilage articular disc separates the joint space into two separate synovial cavities.

Movement
Like other ball and socket joints, movement occurs in all planes, but anteroposterior movement and rotation is slightly restricted. It is involved in the collective movements of the shoulder girdle.

Acromioclavicular Joint

Type of Joint
Synovial plane.

Articulation
Between the lateral end of the clavicle, and the medial border of the acromion of the scapula.

NOTE: A fibrocartilage articular disc partially divides the articular cavity, although it is sometimes absent.

Movement
It is involved in the collective movements of the shoulder girdle, enabling the scapula to change its position in relation to the clavicle.

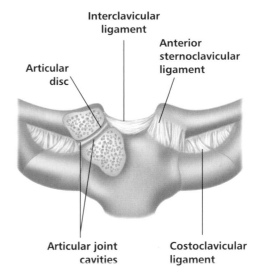

Figure 6.16: The sternoclavicular joint (anterior view). Note: Posterior aspect of joint has a posterior sternoclavicular ligament similar, but weaker, than anterior sternoclavicular ligament.

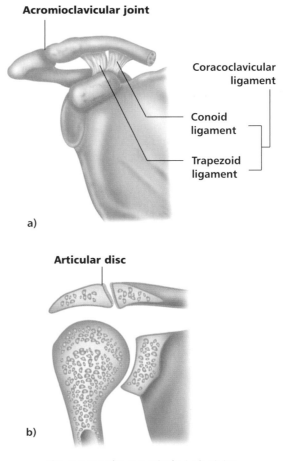

Figure 6.17: The acromioclavicular joint; a) anterior view, b) coronal view.

Shoulder (Glenohumeral) Joint

Type of Joint
Synovial ball and socket.

Articulation
The head of the humerus articulates with the shallow pear shaped glenoid cavity (fossa) of the scapula. This articulation is inherently unstable due to the glenoid cavity being only approximately one-third the size of the humeral head, although it is slightly deepened by a rim of fibrocartilage called the *glenoid labrum* or *labrum glenoidale* (triangular in cross-section). The shoulder joint is the most freely moving joint of the body, precisely because stability has been sacrificed to enable maximum range of movement.

Articular Capsule
Extends from the margin of the glenoid cavity (including part of the labrum), to the anatomical neck of the humerus. The thin capsule is very loose, thus enabling maximum movement of the joint. When the arm is by the side, the lower part of the capsule hangs in a loose fold, which becomes progressively more taut as the arm is abducted; increasingly so if the arm continues into elevation. The capsule contributes very little to the stability of the joint. The surrounding muscles, whose attachments are intimately related to the capsule, largely supply joint stability.

Ligaments
Transverse humeral ligament: Spans the gap between the humeral tubercles. It holds the long head of the biceps brachii in the intertubular sulcus as it leaves the joint.
Glenohumeral ligament: Three slightly thickened bands of longitudinal fibres on the internal surface of the anterior part of the capsule. May be absent.
Coracohumeral ligament: Extends from the coracoid process of the scapula to the upper part of the anatomical neck of the humerus. It greatly reinforces the capsule superiorly and slightly anteriorly.
Coraco-acromial ligament: This ligament is totally unconnected to the articular capsule. It forms a shelf above the joint, running between the coracoid process and the acromion process of the scapula.

Various bursae are associated with the shoulder joint. The most important is the subacromial bursa that separates the coraco-acromial ligament from the supraspinatus tendon located above the shoulder joint.

Stabilizing Tendons
Long head of biceps brachii tendon: Runs from the superior aspect of the glenoid labrum to enter and travel within the joint cavity, thus travelling within the articular capsule (hence it is covered with a sheath of synovial membrane). On leaving the cavity, it enters the intertubular groove of the humerus. Its location secures the head of the humerus tightly against the glenoid cavity, thereby acting as a steadying influence during movements of the shoulder joint.
Rotator cuff tendons: The four rotator cuff tendons (supraspinatus, infraspinatus, teres minor, and subscapularis encircle the joint and fuse with the articular capsule. Consequently, the rotator cuff muscles or tendons are prone to injury if the joint is vigorously circumducted, as in throwing a ball.

NOTE: Because overall, the reinforcements of the shoulder joint are weakest inferiorly, the humerus is more prone to dislocate downwards.

Movements
Flexion, extension, abduction, adduction, medial and lateral rotation, circumduction, plus elevation through flexion and abduction (*see* pp. 14–16, 19).

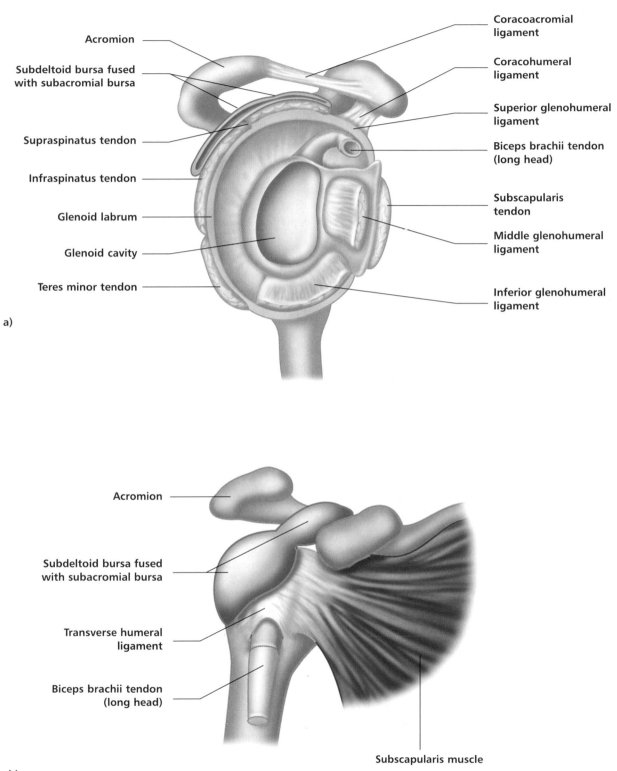

Acromion

Subdeltoid bursa fused with subacromial bursa

Supraspinatus tendon

Infraspinatus tendon

Glenoid labrum

Glenoid cavity

Teres minor tendon

Coracoacromial ligament

Coracohumeral ligament

Superior glenohumeral ligament

Biceps brachii tendon (long head)

Subscapularis tendon

Middle glenohumeral ligament

Inferior glenohumeral ligament

a)

Acromion

Subdeltoid bursa fused with subacromial bursa

Transverse humeral ligament

Biceps brachii tendon (long head)

Subscapularis muscle

b)

Figure 6.18: The shoulder joint; a) right arm, lateral view, b) right arm, anterior view (cut).

Elbow Joint

Type of Joint
Synovial hinge (ginglymus).

Articulation
Upper surface of the head of the radius articulates with the capitulum of the humerus. The trochlear notch of the ulna articulates with the trochlea of the humerus (which constitutes the 'hinge' mechanism and the main stabilizing factor).

Articular Capsule
The relatively loose articular capsule extends from the coronoid and olecranon fossae of the humerus to the coronoid and olecranon processes of the ulna, and to the annular ligament enclosing the head of the radius. The capsule is thin anteriorly and posteriorly to allow flexion and extension, but is strengthened on each side by collateral ligaments.

Ligaments
Ulnar (medial) collateral ligament: Three strong bands reinforcing the medial side of the capsule.
Radial (lateral) collateral ligament: A strong triangular ligament reinforcing the lateral side of the capsule.

Stabilizing Tendons
The tendons of the biceps brachii, triceps brachii, brachialis, plus many muscles located on the forearm: These tendons cross the elbow joint and provide extra security.

Movements
Flexion and extension only.

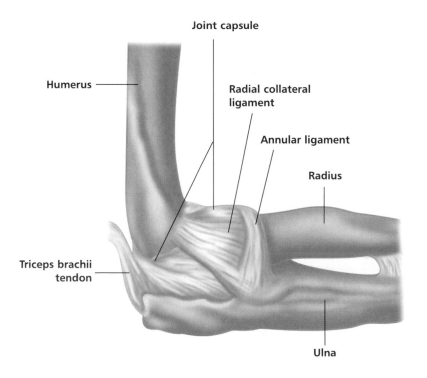

a)

Figure 6.19: The elbow joint; a) right arm, lateral view.

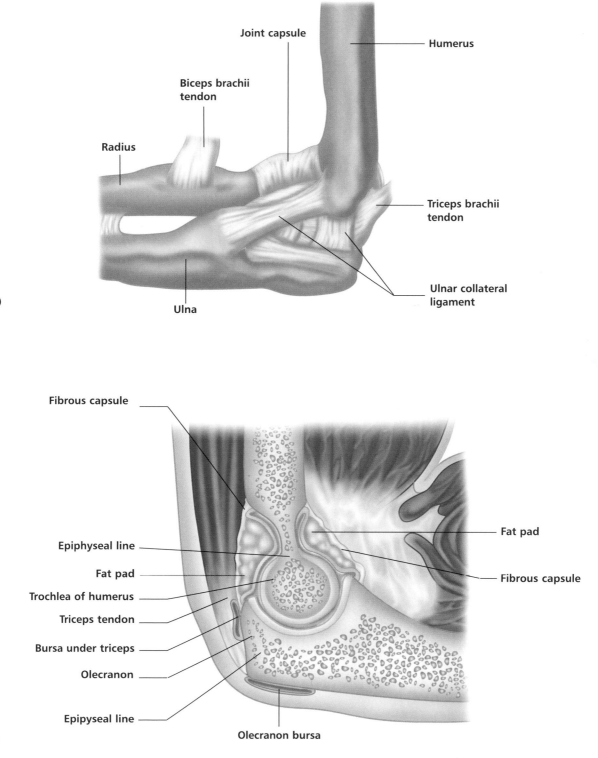

Joint capsule

Humerus

Biceps brachii tendon

Radius

Triceps brachii tendon

Ulnar collateral ligament

Ulna

b)

Fibrous capsule

Epiphyseal line

Fat pad

Fat pad

Trochlea of humerus

Fibrous capsule

Triceps tendon

Bursa under triceps

Olecranon

Epipyseal line

Olecranon bursa

c)

Figure 6.19: The elbow joint; b) right arm, medial view, c) right arm, mid-sagittal view.

Proximal Radio-ulnar Joint

Type of Joint
Synovial pivot.

Articulation
The disc shaped head of the radius rotates within a ring formed by the radial notch on the ulna and the annular ligament of the radius.

NOTE: The synovial cavity of this joint is continuous with that of the elbow joint.

Movements
Pronation and supination of the forearm.

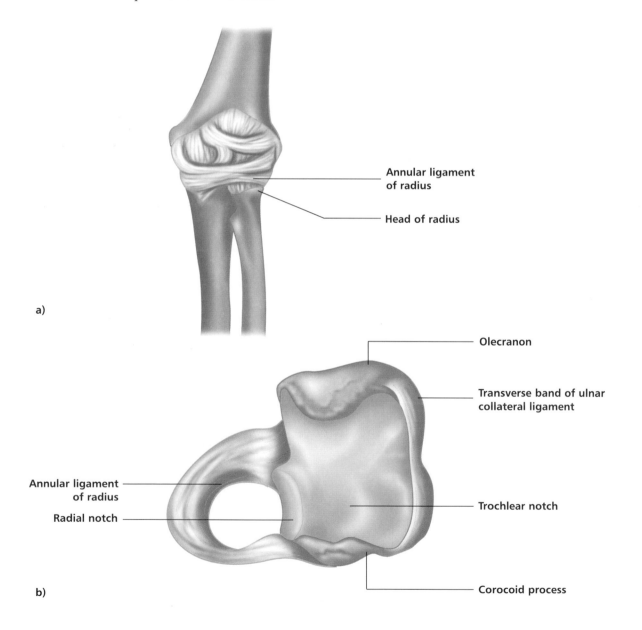

Annular ligament
of radius

Head of radius

a)

Olecranon

Transverse band of ulnar
collateral ligament

Annular ligament
of radius

Radial notch

Trochlear notch

Corocoid process

b)

Figure 6.20: The proximal (superior) radio-ulnar joint; a) left arm, anterior view, b) left arm, superior view.

Distal Radio-ulnar Joint

Type of Joint
Synovial pivot.

Articulation
Between the head of the ulna, and the ulnar notch of the radius.

NOTE: A fibrocartilage articular disc unites the styloid process of the ulna and the medial side of the distal radius.

Movements
Pronation and supination of the forearm.

Intermediate Radio-ulnar Joint

Type of Joint
Syndesmosis.

Articulation
Connects the interosseous border of the radius with the interosseous border of the ulna, via the interosseous membrane. Also, a slender fibrous band called the *oblique cord* connects the ulnar tuberosity to the proximal end of the shaft of the radius.

Function
Increases the surface of origin of the deep forearm muscles; helps bind the radius and ulna together; and transmits to the ulna any force passing upwards from the hand along the radius.

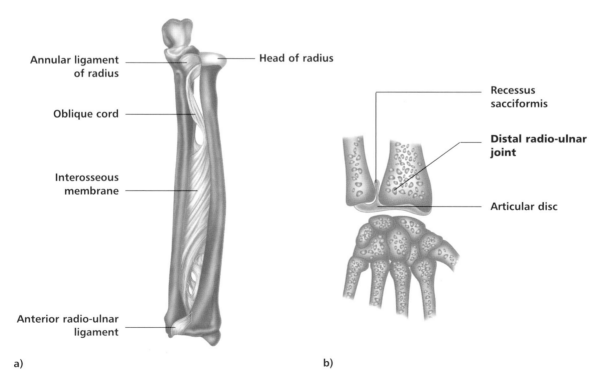

Annular ligament of radius
Head of radius
Oblique cord
Recessus sacciformis
Distal radio-ulnar joint
Interosseous membrane
Articular disc
Anterior radio-ulnar ligament

a)   b)

Figure 6.21: The distal and intermediate radio-ulnar joints; a) left arm, anterior view, b) left arm / hand, coronal view.

Radio-carpal Joint (Wrist Joint)

Type of Joint
Synovial ellipsoid.

Articulation
The distal surface of the radius and the articular disc (the same disc as described with the distal radio-ulnar joint, *see* page 91) articulates with the proximal row of carpals, which are the scaphoid, lunate and triquetral (triquetrum).

Movements
Movements are in combination with the intercarpal joints: flexion, extension, adduction (ulnar deviation), abduction (radial deviation) and circumduction.

Intercarpal Joints

Type of Joint
A series of synovial plane joints.

Articulation
This joint has articulations between the two carpal rows (midcarpal joint), plus articulations between each bone of the proximal carpal row and of the distal carpal row.

Movements
Movements are in combination with the radio-carpal joint: flexion, extension, adduction (ulnar deviation), abduction (radial deviation) and circumduction.

Carpometacarpal Joint of the Thumb

Type of Joint
Synovial saddle joint.

Articulation
Between the trapezium and the base of the first metacarpal bone (the thumb).

Movements
Flexion, extension, abduction, and adduction. At the extreme range of flexion, the first metacarpal medially rotates so that the palmar surface of the thumb becomes opposed to the pads of the fingers. Conversely, slight lateral rotation occurs when the thumb approaches full extension. Combining these movements create approximate circumduction of the thumb.

Common Carpometacarpal Joint

Type of Joint
Synovial plane.

Articulation
Between the distal row of carpal bones and the bases of the medial four metacarpal bones of the hand.

Movements
Very little movement is possible. However, the articulation at the fifth metacarpal with the hamate is a flattened saddle joint, allowing slight opposition of the little finger across the palm.

Intermetacarpal Joints

Type of Joint
Synovial plane.

Articulation
Between adjacent sides of the bases of metacarpal bones 2–5.

Movements
Limited movement between adjacent metacarpals.

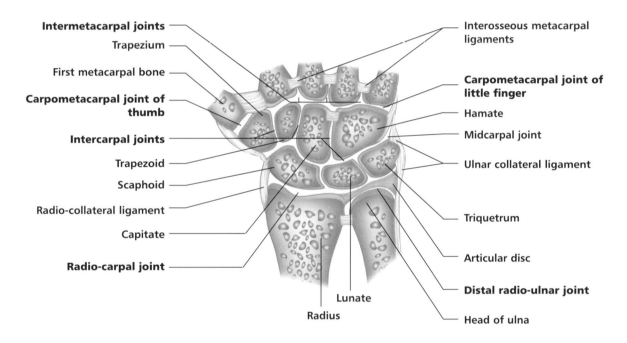

Figure 6.22: The radio-carpal (wrist), intercarpal, carpometacarpal and intermetacarpal joints (coronal view).

Metacarpophalangeal Joints

Type of Joint
Synovial condyloid.

Articulation
Between the head of a metacarpal and the base of a proximal phalanx.

NOTE: The capsule is deficient on the dorsal aspect, where it is replaced by an expansion of the long extensor tendon.

Movements
Flexion and extension. Abduction and adduction (possible only in extension, but with very little movement at the thumb). Combined movements may produce circumduction.

Interphalangeal Joints

Type of Joint
Synovial hinge.

Articulation
Between the proximal and middle phalanges (proximal interphalangeal joint, abbreviated PIP), or the middle and distal phalanges (distal interphalangeal joint, abbreviated DIP).

Movements
Flexion and extension.

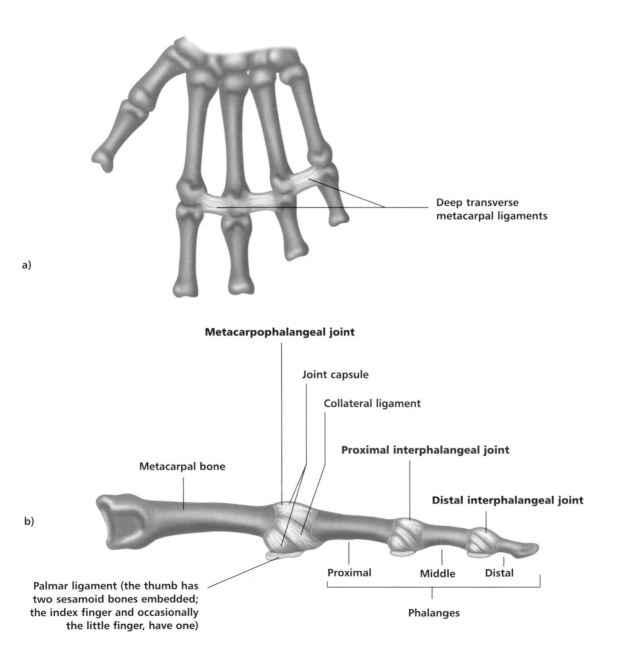

Deep transverse metacarpal ligaments

a)

Metacarpophalangeal joint

Joint capsule

Collateral ligament

Proximal interphalangeal joint

Distal interphalangeal joint

Metacarpal bone

b)

Palmar ligament (the thumb has two sesamoid bones embedded; the index finger and occasionally the little finger, have one)

Proximal   Middle   Distal

Phalanges

Figure 6.23: The metacarpophalangeal and interphalangeal joints; a) anterior view, b) medial view.

Joints of the Pelvic Girdle and Lower Limb

Lumbosacral and Sacrococcygeal Joints

Type of Joint
Both joints: cartilaginous symphysis (slightly movable).

Articulation
Lumbosacral: Between the fifth lumbar vertebra (L5) and the body of the first sacral segment (S1). This joint has the same features as other typical intervertebral joints, with the addition of the iliolumbar ligament.
Sacrococcygeal: Between the last sacral and first coccygeal segments. It is reinforced all round by the sacrococcygeal ligaments.

NOTE: Both joints contain a fibrocartilaginous intervertebral disc.

Movements
The lumbosacral joint contributes to the collective movements of the lumbar vertebral joints. The sacrococcygeal joint has very little functional movement, and is often partially or fully obliterated in old age.

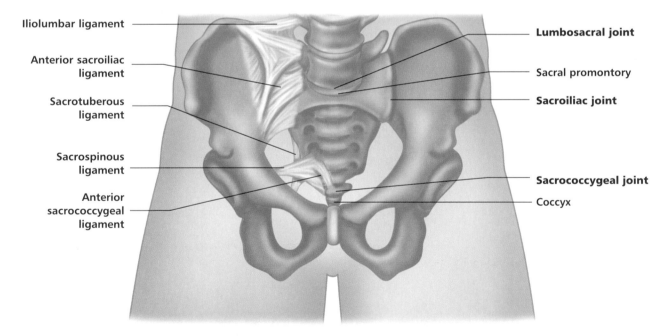

Figure 6.24: The lumbosacral, sacroiliac and sacrococcygeal joints (anterior view).

Sacroiliac Joint

Type of Joint
A synovial joint with pronounced irregular depressions and tubercles on the articular surfaces.

NOTE: The articular surface of the sacrum is hyaline cartilage, but that of the ilium is usually of the fibrous type.

Articulation
Between the auricular surfaces on the sacrum and the iliac bone.

Movements
Very limited movements occur because of the irregular joint surfaces and the strong sacroiliac ligaments.

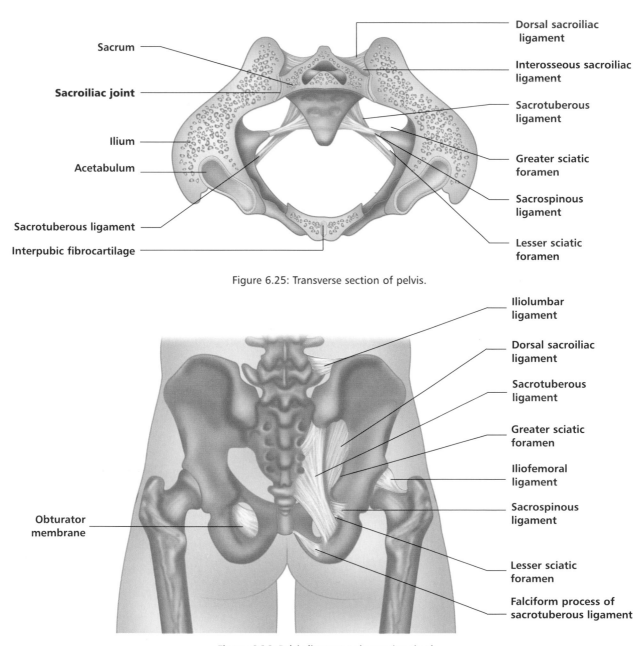

Figure 6.25: Transverse section of pelvis.

Figure 6.26: Pelvic ligaments (posterior view).

Pubic Symphysis

Type of Joint
Cartilaginous symphysis (slightly movable).

Articulation
The midline joint between the superior rami of the pubic bones.

NOTE: The joint contains a fibrocartilaginous interpubic disc with a slit-like cavity, which in women, can develop into a large cavity.

Movements
No significant movement occurs other than some separation of the pubic bones in women during pregnancy and childbirth.

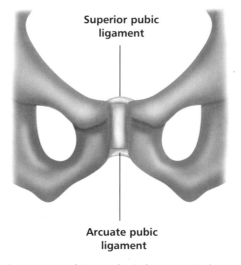

Figure 6.27: Pubic symphysis (anterior view).

Hip Joint

Type of Joint
Synovial ball and socket.

Articulation
The spherical head of the femur articulates with the cup-like acetabulum of the coxal (hip) bone. A circular rim of fibrocartilage called the *acetabular labrum* or *labrum acetabulare,* which grasps the femoral head, enhances the depth of the acetabulum. Unlike the articulation of the shoulder joint, the hip articulation fits securely together.

Articular Capsule
Extends from the rim of the acetabulum to the neck of the femur. It is very strong and tense in extension, which contrasts to the thin and lax capsule of the shoulder joint.

Ligaments
Iliofemoral ligament: A thick and strong triangular band situated anteriorly.
Pubofemoral ligament: A triangular thickening of the inferior aspect of the capsule.
Ischiofemoral ligament: A spiral ligament situated posteriorly.

These three ligaments are arranged so that when a person stands up (i.e. hip joint moves from flexion to extension), the head of the femur is 'screwed' into the acetabulum, and held firmly in position.

Ligament of the head of the femur: Also called the ligamentum teres or the capitate ligament, this flat intracapsular ligament runs from the femoral head to the lower lip of the acetabulum. It contains an artery that is a source of blood for the head of the femur. It is slack during most hip movements and therefore does not contribute to the joint's stability.

Stabilizing Tendons
This joint is inherently stable by virtue of its structure and ligaments. All surrounding muscles and tendons contribute to its stability, but in a very minor capacity compared to those of the shoulder joint.

Movements
Flexion, extension, abduction, adduction, medial and lateral rotation, circumduction (limited, compared to the shoulder joint).

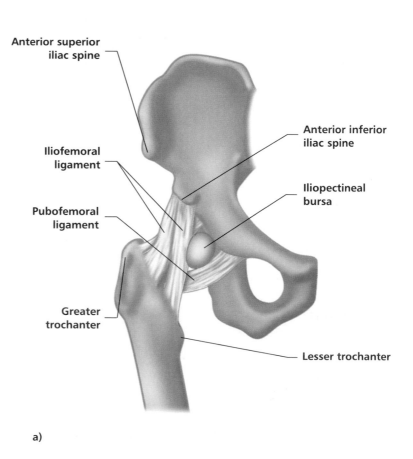

Anterior superior iliac spine

Anterior inferior iliac spine

Iliofemoral ligament

Iliopectineal bursa

Pubofemoral ligament

Greater trochanter

Lesser trochanter

a)

Figure 6.28: The hip joint, a) right leg, anterior view.

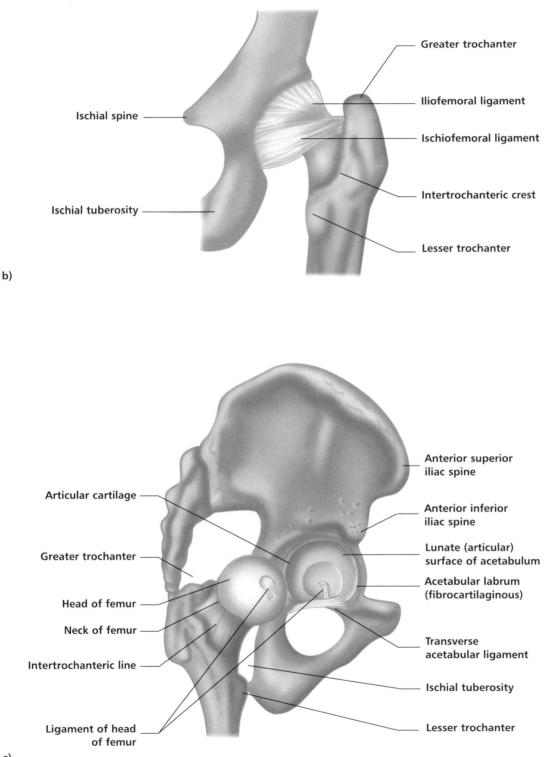

b)

Greater trochanter

Iliofemoral ligament

Ischial spine

Ischiofemoral ligament

Intertrochanteric crest

Ischial tuberosity

Lesser trochanter

c)

Anterior superior iliac spine

Articular cartilage

Anterior inferior iliac spine

Lunate (articular) surface of acetabulum

Greater trochanter

Acetabular labrum (fibrocartilaginous)

Head of femur

Neck of femur

Transverse acetabular ligament

Intertrochanteric line

Ischial tuberosity

Ligament of head of femur

Lesser trochanter

Figure 6.28: The hip joint, b) right leg, posterior view, c) right leg, lateral view.

Knee Joint

The knee joint is the largest and most complex joint in the body. Within its joint cavity it contains three articulations: the lateral and medial articulations of the *tibiofemoral joint*, and the *femoropatellar joint*.

Type of Joint

Tibiofemoral joint: Functionally a modified synovial hinge joint, but structurally a condyloid joint.
Femoropatellar joint: Synovial plane joint.

Articulation

Tibiofemoral joint: Condyles of the femur articulate with the condyles of the tibia; but with two C-shaped menisci or semilunar cartilages between the opposing articular surfaces.
Femoropatellar joint: Posterior surface of patella articulates with patellar surface at the lower end of the femur.

Articular Capsule

The knee is the only joint where the capsule only partially encloses the joint cavity. Instead, the true capsular fibres are integrated within a ligamentous sheath composed of muscle tendons or expansions from them, which collectively encapsulate the joint. True capsular fibres are located only at the sides and posterior of the joint.

Extracapsular (extra-articular) Ligaments

Tibial (medial) collateral ligament: A broad, flat band running from the medial epicondyle of the femur, downwards and forwards to the medial condyle of the tibial shaft. Some fibres are fused to the medial meniscus.
Fibular (lateral) collateral ligament: A round, cord-like ligament, fully detached from the thin lateral part of the capsule. It extends from the lateral epicondyle of the femur, downwards and backwards to the head of the fibula.
Oblique popliteal ligament: An expansion of the semimembranosus tendon, that passes upward and laterally over the posterior of the joint.
Arcuate popliteal ligament: Extends from the head of the fibula upwards and medially, spreading into the back of the capsule and to the lateral condyle of the femur; thus reinforcing the back of the joint.

Intracapsular (intra-articular) Ligaments and Menisci

Anterior cruciate ligament: Extends obliquely upwards, laterally and backwards from the anterior intercondylar area of the tibia to the medial surface of lateral femoral condyle. It prevents posterior displacement of the femur on the tibia, and also helps check hyperextension of the knee.
Posterior cruciate ligament: Passes upwards, medially and forwards from the posterior intercondylar area of the tibia to the lateral side of the medial femoral condyle. Thus it lies on the medial side of the weaker anterior cruciate. It prevents anterior displacement of the femur on the tibia.

The *cruciate ligaments* are within the joint capsule, but outside the joint cavity. Synovial membrane covers most of their surface.

Menisci: Between the femoral and tibial condyles are two crescent shaped fibrous wedges called menisci that help compensate for the incongruence of the articular surfaces. They also help absorb shock transmitted to the knee joint. The menisci are attached only at their outer margins and are prone to tearing. The medial meniscus is also attached to the tibial collateral ligament, and is therefore much more firmly anchored than the lateral meniscus, which does not attach to the fibular collateral ligament.
Medial and lateral coronary ligaments: Capsular fibres that attach the menisci to the tibial condyles.
Transverse ligament of the knee: A fibrous band that joins the anterior parts of the menisci.

Stabilizing Tendons

Patellar ligament (ligamentum patellae): This strong ligament is actually the distal part of the quadriceps tendon. It runs from the patella (which is embedded within the tendon as a sesamoid bone – *see* page 33) to the tibial tuberosity. Other thinner bands called the *medial and lateral patellar retinacula* pass down the sides of the patella to attach to the front of each tibial condyle; effectively substituting for the capsule anteriorly. *Tendon of semimembranosus*: Helps reinforce the posterior of the knee joint.

The muscles surrounding the knee joint are particularly crucial as stabilizers.

Movements

Flexion, extension. Some rotation can occur when the knee is flexed. Also, as a result of the tightening of various ligaments (especially the cruciates) and tendons, slight medial rotation of the femur occurs upon the fixed tibia as the knee straightens into full extension. (When both the femur and tibia are not fixed, as in kicking, the tibia rotates laterally at the end of extension and medially at the beginning of flexion).

NOTE: The popliteus muscle 'unlocks' the extended knee joint prior to flexion, enabling flexion to occur.

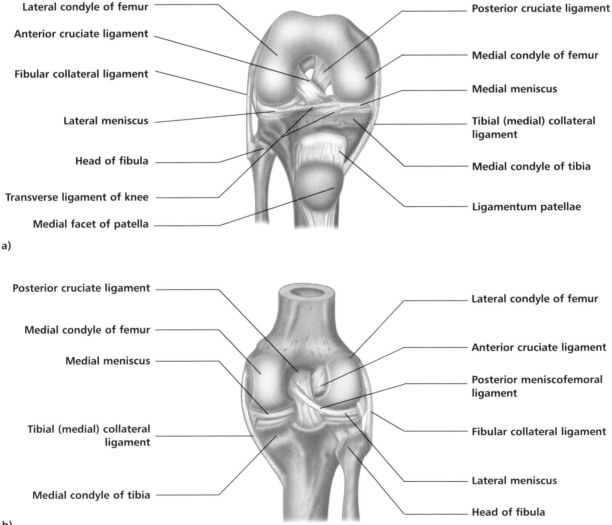

Figure 6.29: The knee joint; a) right leg, anterior view, b) right leg, posterior view.

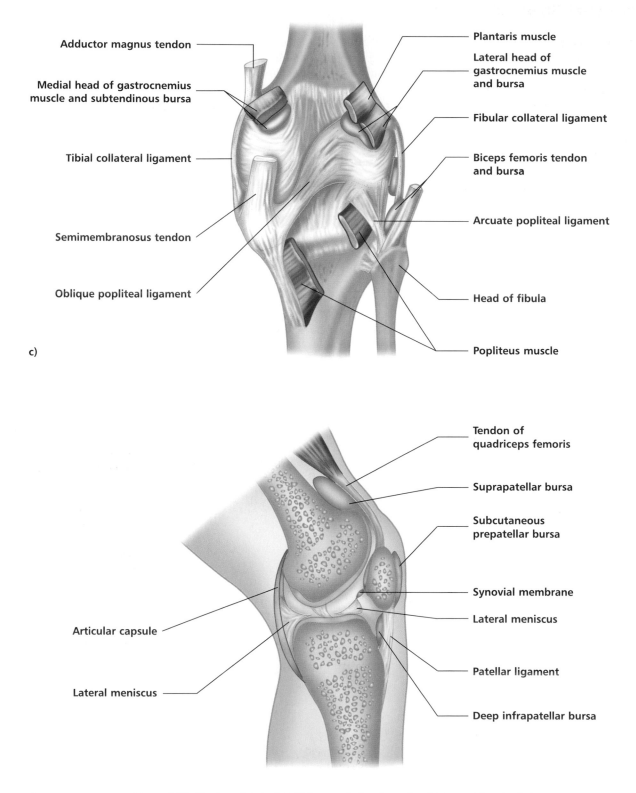

Adductor magnus tendon

Medial head of gastrocnemius muscle and subtendinous bursa

Tibial collateral ligament

Semimembranosus tendon

Oblique popliteal ligament

Plantaris muscle

Lateral head of gastrocnemius muscle and bursa

Fibular collateral ligament

Biceps femoris tendon and bursa

Arcuate popliteal ligament

Head of fibula

Popliteus muscle

c)

Tendon of quadriceps femoris

Suprapatellar bursa

Subcutaneous prepatellar bursa

Synovial membrane

Lateral meniscus

Patellar ligament

Deep infrapatellar bursa

Articular capsule

Lateral meniscus

d)

Figure 6.29: The knee joint; c) right leg, posterior view, d) right leg, mid-sagittal view.

Proximal Tibiofibular Joint

Type of Joint
Synovial plane.

Articulation
Between a facet on the head of the fibula and a similar facet on the lateral condyle of the tibia.

Movements
Movement is slight and passively occurs along with movements of the ankle joint.

Distal Tibiofibular Joint

Type of Joint
Syndesmosis.

Articulation
Between the rough, triangular, opposed surfaces at the distal end of the tibia and fibula.

Movements
Movement is slight and passively occurs along with movements of the ankle joint.

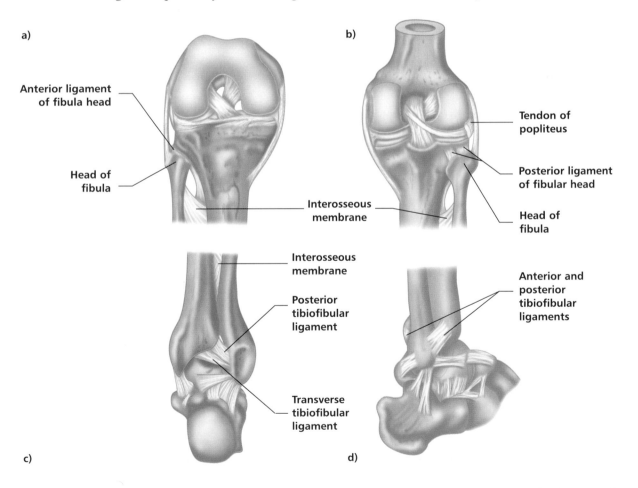

a)

Anterior ligament of fibula head

Head of fibula

b)

Tendon of popliteus

Posterior ligament of fibular head

Head of fibula

Interosseous membrane

Interosseous membrane

Posterior tibiofibular ligament

Transverse tibiofibular ligament

Anterior and posterior tibiofibular ligaments

c)

d)

Figure 6.30: The tibiofibular joints; a) proximal tibiofibular joint, right leg, anterior view, b) proximal tibiofibular joint, right leg, posterior view, c) distal tibiofibular joint, right leg, posterior view, d) distal tibiofibular joint, right leg, lateral view.

Ankle Joint

Type of Joint
Synovial hinge.

Articulation
Between the distal tibia, the medial malleolus of the tibia, the lateral malleolus of the fibula and the talus. Therefore, the lower ends of the tibia and fibula provide a socket for the talus.

Movements
Dorsiflexion and plantar flexion.

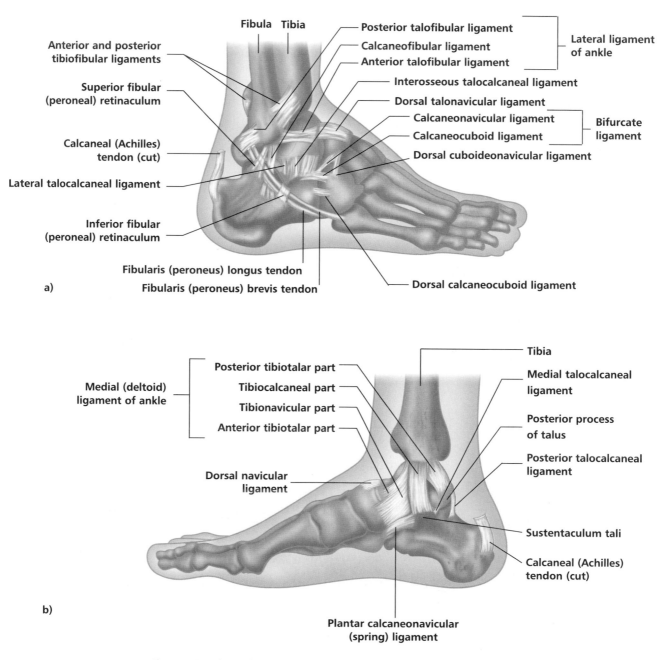

Figure 6.31: The ankle joint; a) right foot, lateral view, b) right foot, medial view.

The Arches of the Foot

The longitudinal arch:

- A series of synovial plane joints.
- It extends from the calcaneus to the metatarsals via the talus, navicular and cuneiforms.
- The shapes of the metatarsal bones form it.
- The calcaneonavicular (spring) ligament, a number of small interosseous ligaments, and the tendons of the tibialis anterior and tibialis posterior muscles support it.
- The arch is higher on the medial side than the lateral side.

The transverse arch:

- A series of synovial plane joints.
- It is placed through the distal row of tarsal bones.
- The shape of the tarsal bones, many small interosseous ligaments and the tendons of the peroneus longus, tibialis anterior and tibialis posterior muscles support it.

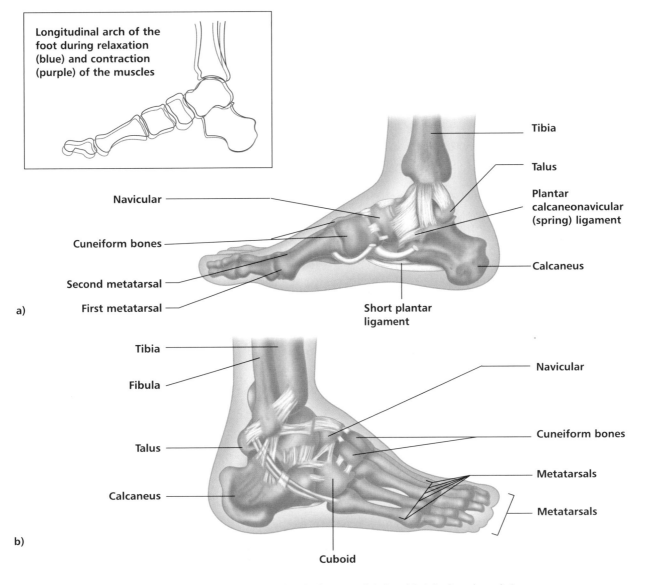

Figure 6.32: The arches of the foot; a) right foot, medial view, b) right foot, lateral view.

Intertarsal Joints

Type of Joints
A complex set of synovial plane joints.

Articulation
Subtalar joint: Between the inferior surface of the talus and the superior surface of the calcaneus.
Talocalcaneonavicular joint: Between the talus, calcaneus and navicular.
Calcaneocuboid joint: Between the calcaneus and cuboid.
Transverse tarsal joint: A term to describe the transverse plane extending across the full width of the tarsus, comprising the talocalcaneonavicular joint and the calcaneocuboid joint.
Cuneonavicular joint: Between the cuneiform and the navicular.
Intercuneiform joints: Between the three cuneiform bones.
Cuneocuboid joint: Between the lateral cuneiform bone and the cuboid bone.

Movements of the Tarsus
Inversion and eversion of the foot.

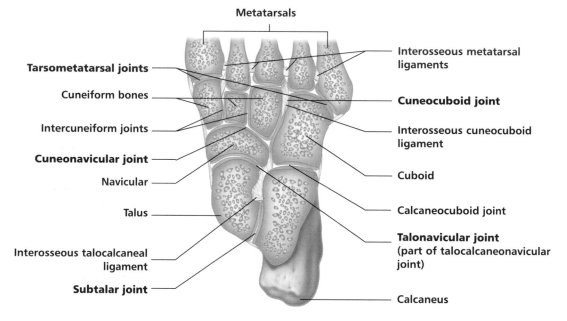

Figure 6.33: The intertarsal joints (horizontal section of right foot).

Tarsometatarsal and Intermetatarsal Joints

Type of Joints
Synovial plane.

Articulation
Tarsometatarsal joints: Between the distal (anterior) row of tarsal bones (the cuboid and three cuneiforms) and the bases of the metatarsal bones.
Intermetatarsal joints: Between facets on adjacent sides of the bases of all lateral metatarsal bones.

Movements
Small gliding movements of the metatarsals, limited by ligaments and the interlocking of the bones, contribute slightly to inversion and eversion of the foot.

Metatarsophalangeal Joints

Type of Joint
Synovial condyloid.

Articulation
Between the head of a metatarsal and the base of a proximal phalanx.

NOTE: The capsule is deficient on the dorsal aspect, where it is replaced by an expansion of the extensor tendon.

Movements
Flexion and extension. Abduction and adduction. Combined movements may produce passive circumduction.

NOTE: In flexion, the toes are drawn together; in extension they tend to spread apart and incline slightly laterally. Movements are less extensive than at the corresponding joints of the hand.

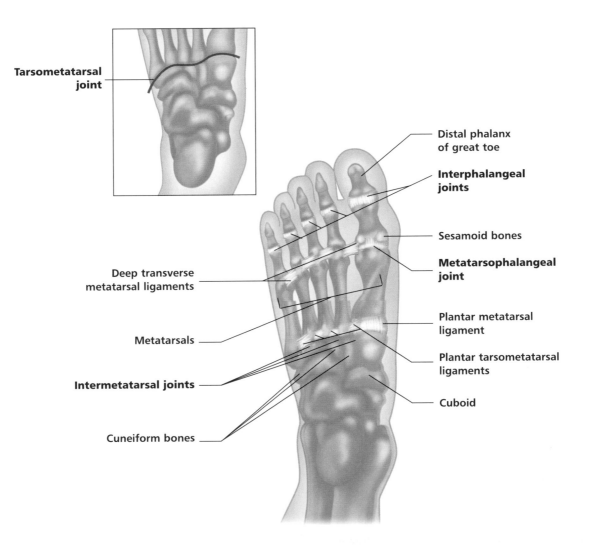

Figure 6.34: The tarsometatarsal, intermetatarsal and metatarsophalangeal joints (plantar view).

Interphalangeal Joints

Type of Joint
Synovial hinge.

Articulation
Between the proximal and middle phalanges (proximal interphalangeal joint, abbreviated PIP), or the middle and distal phalanges (distal interphalangeal joint, abbreviated DIP).

Movements
Flexion and extension.

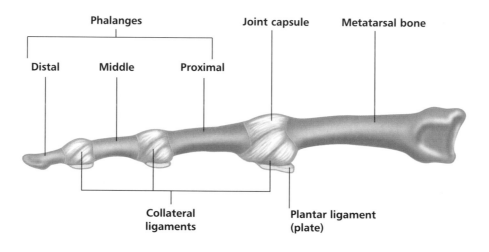

Figure 6.35: The metatarsophalangeal and interphalangeal joints (lateral view).

	Movement	Muscle
Mandible	Elevation	Temporalis (anterior fibres); Masseter; Pterygoideus Medialis
	Depression	Pterygoideus Lateralis; Digastricus; Mylohyoideus; Geniohyoideus
	Protraction	Pterygoideus Lateralis; Pterygoideus Medialis; Masseter (superficial fibres)
	Retraction	Temporalis (horizontal fibres); Digastricus
	Chewing	Pterygoideus Lateralis; Pterygoideus Medialis; Masseter; Temporalis
Larynx	Elevation	Digastricus; Stylohyoideus; Mylohyoideus; Geniohyoideus; Thyrohyoideus
	Depression	Sternohyoideus; Sternothyroideus; Omohyoideus
	Protraction	Geniohyoideus
	Retraction	Stylohyoideus
Atlanto-occipital & Atlanto-axial Joints	Flexion	Longus Capitis; Rectus Capitis Anterior; Sternocleidomastoideus (anterior fibres)
	Extension	Semispinalis Capitis; Splenius Capitis; Rectus Capitis Posterior Major; Rectus Capitis Posterior Minor; Obliquus Capitis Superior; Longissimus Capitis; Trapezius; Sternocleidomastoideus (posterior fibres)
	Rotation and Lateral Flexion	Sternocleidomastoideus; Obliquus Capitis Inferior; Obliquus Capitis Superior; Rectus Capitis Lateralis; Longissimus Capitis; Splenius Capitis
Intervertebral Joints Cervical Region	Flexion	Longus Colli; Longus Capitis; Sternocleidomastoideus
	Extension	Longissimus Cervicis; Longissimus Capitis; Splenius Capitis; Splenius Cervicis; Semispinalis Cervicis; Semispinalis Capitis; Trapezius; Interspinales; Iliocostalis Cervicis
	Rotation and Lateral Flexion	Longissimus Cervicis; Longissimus Capitis; Splenius Capitis; Splenius Cervicis; Multifidis; Longus Colli; Scalenus Anterior; Scalenus Medius; Scalenus Posterior; Sternocleidomastoideus; Levator Scapulae; Iliocostalis Cervicis; Intertransversarii
Intervertebral Joints Thoracic/Lumbar Regions	Flexion	Muscles of Anterior Abdominal Wall
	Extension	Erector Spinae; Quadratus Lumborum; Trapezius
	Rotation and Lateral Flexion	Iliocostalis Lumborum; Iliocostalis Thoracis; Multifidis; Rotatores; Intertransversarii; Iliocostalis Lumborum; Iliocostalis Thoracis; Multifidis; Rotatores; Intertransversarii; Quadratus Lumborum; Psoas Major; Muscles of Anterior Abdominal Wall

	Movement	Muscle
Shoulder Girdle	Elevation	Trapezius (upper fibres); Levator Scapulae; Rhomboideus Minor; Rhomboideus Major; Sternocleidomastoideus
	Depression	Trapezius (lower fibres); Pectoralis Minor; Pectoralis Major (sternocostal portion); Latissimus Dorsi
	Protraction	Serratus Anterior; Pectoralis Minor; Pectoralis Major
	Retraction	Trapezius (middle fibres); Rhomboideus Minor; Rhomboideus Major; Latissimus Dorsi
	Lateral Displacement of Inferior Angle of Scapula	Serratus Amterior; Trapezius (upper and lower fibres)
	Medial Displacement of Inferior Angle of Scapula	Pectoralis Minor; Rhomboideus Minor; Rhomboideus Major; Latissimus Dorsi
Shoulder Joint	Flexion	Deltoideus (anterior portion); Pectoralis Major (clavicular portion : sternocostal portion flexes the extended humerus as far as the position of rest); Biceps Brachii; Coracobrachialis
	Extension	Deltoideus (posterior portion); Teres Major (of flexed humerus); Latissimus Dorsi (of flexed humerus); Pectoralis Major (sternocostal portion of flexed humerus); Triceps Brachii (long head to position of rest)
	Abduction	Deltoideus (middle portion); Supraspinatus; Biceps Brachii (long head)
	Adduction	Pectoralis Major; Teres Major; Latissimus Dorsi; Triceps Brachii (long head); Coracobrachialis
	Lateral Rotation	Deltoideus (posterior portion); Infraspinatus; Teres Minor
	Medial Rotation	Pectoralis Major; Teres Major; Latissimus Dorsi; Deltoideus (anterior portion); Subscapularis
	Horizontal Flexion	Deltoideus (anterior portion); Pectoralis Major; Subscapularis
	Horizontal Extension	Deltoideus (posterior portion); Infraspinatus
Elbow Joint	Flexion	Brachialis; Biceps Brachii; Brachioradialis; Extensor Carpi Radialis Longus; Pronator Teres; Flexor Carpi Radialis
	Extension	Triceps Brachii; Anconeus
Radio-ulnar Joints	Supination	Supinator; Biceps Brachii; Extensor Pollicis Longus
	Pronation	Pronator Quadratus; Pronator Teres; Flexor Carpi Radialis

Movement	Muscle
Radio-carpal and Midcarpal Joints	
Flexion	Flexor Carpi Radialis; Flexor Carpi Ulnaris; Palmaris Longus; Flexor Digitorum Superficialis; Flexor Digitorum Profundus; Flexor Pollicis Longus; Abductor Pollicis Longus; Extensor Pollicis Brevis
Extension	Extensor Carpi Radialis Brevis; Extensor Carpi Radialis Longus; Extensor Carpi Ulnaris; Extensor Digitorum; Extensor Indicis; Extensor Pollicis Longus; Extensor Digiti Minimi
Abduction	Extensor Carpi Radialis Brevis; Extensor Carpi Radialis Longus; Flexor Carpi Radialis; Abductor Pollicis Longus; Extensor Pollicis Longus; Extensor Pollicis Brevis
Adduction	Flexor Carpi Ulnaris; Extensor Carpi Ulnaris
Metacarpophalangeal Joints of the Fingers	
Flexion	Flexor Digitorum Profundus; Flexor Digitorum Superficialis; Lumbricales; Interossei; Flexor Digiti Minimi; Abductor Digiti Minimi; Palmaris Longus (through palmar aponeurosis)
Extension	Extensor Digitorum; Extensor Indicis; Extensor Digiti Minimi
Abduction and Adduction	Interossei; Abductor Digiti Minimi; Lumbricales (may assist in radial deviation); Extensor Digitorum (abducts by hyperextending; tendon to index radially deviates); Flexor Digitorum Profundus (adducts by flexing); Flexor Digitorum Superficialis (adducts by flexing)
Rotation	Lumbricales; Interossei (movement slight except index; only effective when phalanx is flexed); Opponens Digiti Minimi (rotates little finger at carpometacarpal joint)
Interphalangeal Joints of the Fingers	
Flexion	Flexor Digitorum Profundus (both joints); Flexor Digitorum Superficialis (proximal joint only)
Extension	Extensor Digitorum; Extensor Digiti Minimi; Extensor Indicis; Lumbricales; Interossei
Carpometacarpal Joint of the Thumb	
Flexion	Flexor Pollicis Brevis; Flexor Pollicis Longus; Opponens Pollicis
Extension	Extensor Pollicis Brevis; Extensor Pollicis Longus; Abductor Pollicis Longus
Abduction	Abductor Pollicis Brevis; Abductor Pollicis Longus
Adduction	Adductor Pollicis; Dorsal Interossei (first only); Extensor Pollicis Longus (in full extension / abduction); Flexor Pollicis Longus (in full extension / abduction)
Opposition	Opponens Pollicis; Abductor Pollicis Brevis; Flexor Pollicis Brevis; Flexor Pollicis Longus; Adductor Pollicis

	Movement	Muscle
Metacarpophalangeal Joint of the Thumb	Flexion	Flexor Pollicis Brevis; Flexor Pollicis Longus; Palmar Interossei (first only); Abductor Pollicis Brevis
	Extension	Extensor Pollicis Brevis; Extensor Pollicis Longus
	Abduction	Abductor Pollicis Brevis
	Adduction	Adductor Pollicis; Palmar Interossei (first only)
Interphalangeal Joint of the Thumb	Flexion	Flexor Pollicis Longus
	Extension	Abductor Pollicis Brevis; Extensor Pollicis Longus; Adductor Pollicis; Extensor Pollicis Brevis (occasional insertion)
Hip Joint	Flexion	Iliopsoas; Rectus Femoris; Tensor Fasciae Latae; Sartorius; Adductor Brevis; Adductor Longus; Pectineus
	Extension	Gluteus Maximus; Semitendinosus; Semimembranosus; Biceps Femoris (long head); Adductor Magnus (ischial fibres)
	Abduction	Gluteus Medius; Gluteus Minimus; Tensor Fasciae Latae; Obturator Internus (in flexion); Piriformis (in flexion)
	Adduction	Adductor Magnus; Adductor Brevis; Adductor Longus; Pectineus; Gracilis; Gluteus Maximus (lower fibres); Quadratus Femoris
	Lateral Rotation	Gluteus Maximus; Obturator Internus; Gemelli; Obturator Externus; Quadratus Femoris; Piriformis; Sartorius; Adductor Magnus; Adductor Brevis; Adductor Longus
	Medial Rotation	Iliopsoas (in initial stage of flexion); Tensor Fasciae Latae; Gluteus Medius (anterior fibres); Gluteus Minimus (anterior fibres)
Knee Joint	Flexion	Semitendinosus; Semimembranosus; Biceps Femoris; Gastrocnemius; Plantaris; Sartorius; Gracilis; Popliteus
	Extension	Quadratus Femoris
	Medial Rotation of Tibia on Femur	Popliteus; Semitendinosus; Semimembranosus; Sartorius; Gracilis
	Lateral Rotation of Tibia on Femur	Biceps Femoris
Ankle Joint	Dorsiflexion	Tibialis Anterior, Extensor Hallucis Longus; Extensor Digitorum Longus; Fibularis (Peroneus) Tertius
	Plantar Flexion	Gastrocnemius; Plantaris; Soleus; Tibialis Posterior; Flexor Hallucis Longus; Flexor Digitorum Longus; Fibularis (Peroneus) Longus; Fibularis (Peroneus) Brevis

	Movement	Muscle
Intertarsal Joints	Inversion	Tibialis Anterior; Tibialis Posterior
	Eversion	Fibularis (Peroneus) Tertius; Fibularis (Peroneus) Longus; Fibularis (Peroneus) Brevis
	Other Movements	Sliding movements which allow some dorsiflexion, plantar flexion, abduction and adduction, are produced by the muscles acting on the toes. Tibialis Anterior, Tibialis Posterior, and Fibularis (Peroneus) Tertius are also involved.
Metatarsophalangeal Joints of the Toes	Flexion	Flexor Hallucis Brevis; Flexor Hallucis Longus; Flexor Digitorum Longus; Flexor Digitorum Brevis; Flexor Digiti Minimi Brevis; Lumbricales; Interossei
	Extension	Extensor Hallucis Longus; Extensor Digitorum Brevis; Extensor Digitorum Longus
	Abduction and Adduction	Abductor Hallucis; Adductor Hallucis; Interossei; Abductor Digiti Minimi
Interphalangeal Joints of the Toes	Flexion	Flexor Hallucis Longus; Flexor Digitorum Brevis (proximal joint only); Flexor Digitorum Longus
	Extension	Extensor Hallucis Longus; Extensor Digitorum Brevis (not in great toe); Extensor Digitorum Longus; Lumbricales

Skeletal Muscle and Fascia

7

Skeletal Muscle
Structure and
Function

Musculo-skeletal
Mechanics

Skeletal Muscle Structure and Function

Skeletal (somatic or voluntary) muscles make up approximately 40% of the total human body weight. Their primary function is to produce movement through the ability to contract and relax in a coordinated manner. They are attached to bone by tendons. The place where a muscle attaches to a relatively stationary point on a bone, either directly or via a tendon, is called the *origin*. When the muscle contracts, it transmits tension to the bones across one or more joints, and movement occurs. The end of the muscle that attaches to the bone that moves is called the *insertion*.

Overview of Skeletal Muscle Structure

The functional unit of skeletal muscle is known as a *muscle fibre*, which is an elongated, cylindrical cell with multiple nuclei, ranging from 10 to 100 microns in width, and a few millimetres to 30+ centimetres in length. The cytoplasm of the fibre is called the *sarcoplasm*, which is encapsulated inside a cell membrane called the *sarcolemma*. A delicate membrane known as the *endomysium* surrounds each individual fibre.

These fibres are grouped together in bundles covered by the *perimysium*. These bundles are themselves grouped together, and the whole muscle is encased in a sheath called the *epimysium*. These muscle membranes lie through the entire length of the muscle, from the tendon of origin to the tendon of insertion. This whole structure is sometimes referred to as the *musculo-tendinous unit*.

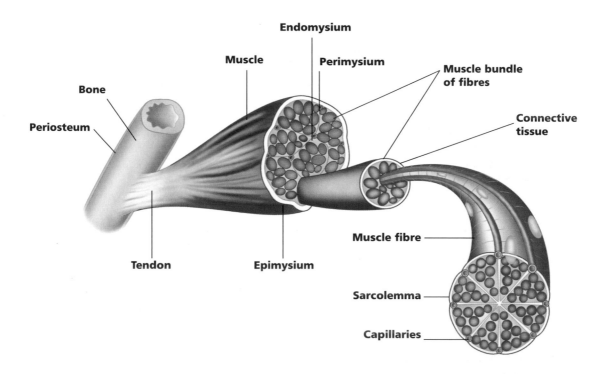

Figure 7.1: Cross-section of muscle tissue.

So, in defining the structure of muscle tissue in more detail, from the minute to gross, we have the following components:

Myofibrils

Through an electron microscope, one can distinguish the contractile elements of a muscle fibre, known as myofibrils, running the whole length of the fibre. Each myofibril reveals alternate light and dark banding, producing the characteristic cross-striation of the muscle fibre. These bands are called *myofilaments*. The light bands are referred to as isotropic (I) bands, and consist of thin myofilaments made of the protein actin. The dark bands are called anisotropic (A) bands, consisting of thicker myofilaments made of the protein myosin. (Note that a third connecting filament made of a protein called titin is now recognized). The myosin filaments have paddle-like extensions that emanate from the filaments rather like the oars of a boat. These extensions latch on to the actin filaments, forming what are described as 'cross-bridges' between the two types of filaments. The cross-bridges, using the energy of ATP, pull the actin strands closer together*. Thus, the light and dark sets of filaments increasingly overlap, like the interlocking of fingers, resulting in muscle contraction. One set of actin-myosin filaments is called a *sarcomere*.

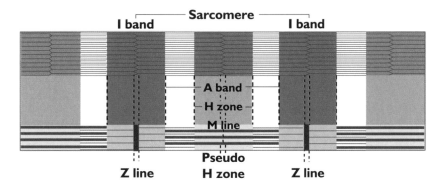

Figure 7.2: The myofilaments in a sarcomere. A sarcomere is bounded at both ends by the Z line.

- The lighter zone is known as the I band, and the darker zone the A band.
- The Z line is a thin dark line at the midpoint of the I band.
- A sarcomere is defined as the section of myofibril between one Z line and the next.
- The centre of the A band contains the H zone.
- The M line bisects the H zone, and delineates the centre of the sarcomere.

If an outside force causes a muscle to stretch beyond its resting level of tonus, the interlinking effect of the actin and myosin filaments that occurs during contraction is reversed. Initially, the actin and myosin filaments accommodate the stretch, but as the stretch continues, the titin filaments increasingly 'pay out' to absorb the displacement. Thus, it is the titin filament that determines the muscle fibre's extensibility and resistance to stretch. Research indicates that a muscle fibre (sarcomere), if properly prepared, can be elongated up to 150% of its normal length at rest.

Endomysium

A delicate connective tissue called endomysium lies outside the sarcolemma of each muscle fibre, separating each fibre from its neighbours, but also connecting them together.

* Huxley's Sliding Filament Theory

The generally accepted hypothesis to explain muscle function is partly described by Huxley's sliding filament theory (Huxley and Hanson, 1954). Muscle fibres receive a nerve impulse that cause the release of calcium ions stored in the muscle. In the presence of the muscles fuel, known as adenosine triphosphate (ATP), the calcium ions bind with the actin and myosin filaments to form an electrostatic (magnetic) bond. This bond causes the fibres to shorten, resulting in their contraction or increase in tonus. When the nerve impulse ceases, the muscle fibres relax. Their elastic elements recoil the filaments to their non-contracted lengths, i.e. their resting level of tonus.

Fasciculi
Muscle fibres are arranged in parallel bundles called fasciculi.

Perimysium
Each fasciculus is bound by a denser collagenic sheath called the perimysium.

Epimysium
The entire muscle, which is therefore an assembly of fasciculi, is wrapped in a fibrous sheath called the epimysium.

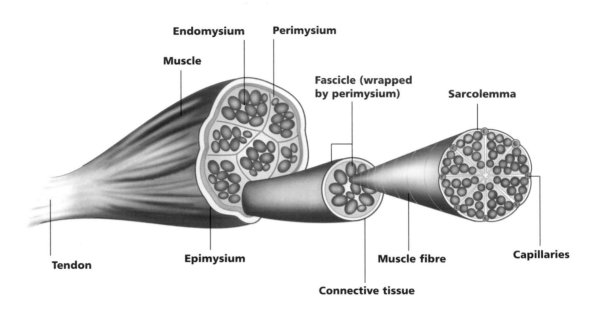

Figure 7.3: The connective tissue sheaths of skeletal muscle.

Deep Fascia
A coarser sheet of fibrous connective tissue lies outside the epimysium, binding individual muscles into functional groups. This deep fascia extends to wrap around other adjacent structures.

Muscle Attachment
The way a muscle attaches to bone or other tissues is either through a direct attachment or an indirect attachment. A direct attachment (called a fleshy attachment) is where the perimysium and epimysium of the muscle unite and fuse with the periosteum of a bone, perichondrium of a cartilage, a joint capsule, or the connective tissue underlying the skin (some muscles of facial expression being good examples of the latter). An indirect attachment is where the connective tissue components of a muscle fuse together into bundles of collagen fibres to form an intervening tendon. Indirect attachments are much more common. The types of tendinous attachments are as follows:

Tendons and Aponeurosis
Muscle fascia, which is the connective tissue component of a muscle, combine together and extend beyond the end of the muscle as round cords or flat bands, called tendons; or as a thin, flat and broad aponeurosis. The tendon or aponeurosis secures the muscle to the bone or cartilage, to the fascia of other muscles, or to a seam of fibrous tissue called a *raphe*. Flat patches of tendon may form on the body of a muscle where it is exposed to friction. This may occur for example on the deep surface of trapezius, where it rubs against the spine of the scapula.

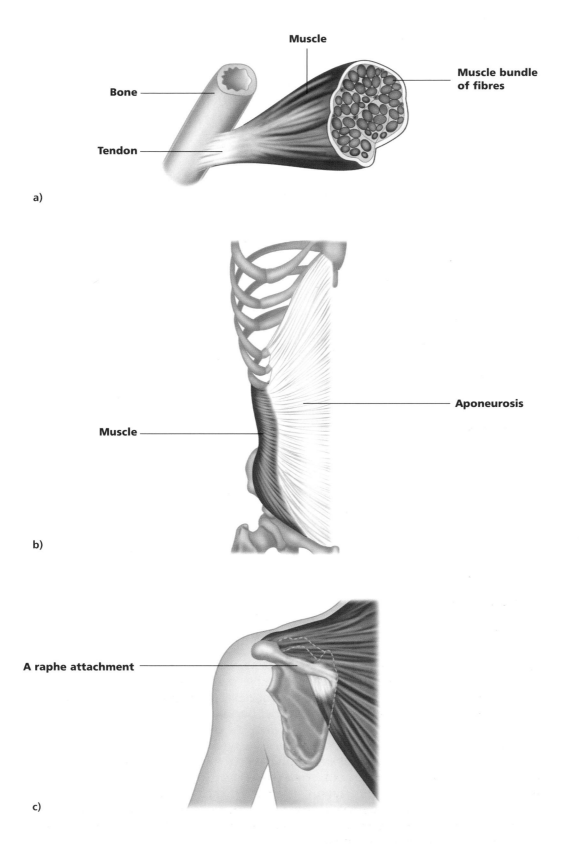

Figure 7.4: a) a tendon attachment, b) attachment by aponeurosis,
c) flat patches of tendon on the deep surface of trapezius.

Intermuscular Septa

In some cases, flat sheets of dense connective tissue known as intermuscular septa penetrate between muscles, providing another medium to which muscle fibres may attach.

Sesamoid Bones

If a tendon is subject to friction, it may, but not necessarily, develop a sesamoid bone within its substance. An example is the peroneus longus tendon in the sole of the foot. However, sesamoid bones may also appear in tendons not subject to friction.

Multiple Attachments

Many muscles have only two attachments, one at each end. However, more complex muscles are often attached to several different structures at its origin and / or its insertion. If these attachments are separated, effectively meaning the muscle gives rise to two or more tendons and/or aponeurosis inserting into different places, the muscle is said to have two heads. For example, the biceps brachii has two heads at its origin: one from the corocoid process of the scapula and the other from the supraglenoid tubercle. The triceps brachii has three heads and the quadriceps has four.

Red and White Muscle Fibres

There are three types of skeletal muscle fibres; red slow-twitch fibres, white fast-twitch fibres, and intermediate fast-twitch fibres.

1. Red Slow-twitch Fibres

These are thin cells that contract slowly. The red colour is due to their content of myoglobin, a substance similar to haemoglobin, which stores oxygen and increases the rate of oxygen diffusion within the muscle fibre. As long as oxygen supply is plentiful, red fibres can contract for sustained periods, and are thus very resistant to fatigue. Successful marathon runners tend to have a high percentage of these red fibres.

2. White Fast-twitch Fibres

These are large cells that contract rapidly. They are pale, due to a lesser content of myoglobin. They fatigue quickly, because they rely on short-lived glycogen reserves in the fibre to contract. However, they are capable of generating much more powerful contractions than red fibres, enabling them to perform rapid, powerful movements for short periods. Successful sprinters have a higher proportion of these white fibres.

3. Intermediate Fast-twitch Fibres

These red or pink fibres are a compromise in size and activity between the red and white fibres.

NOTE: There is always a mixture of these muscle fibres in any given muscle, giving them a range of fatigue resistance and contractile speeds.

Blood Supply

In general, each muscle receives one artery to *bring* nutrients via blood into the muscle, and several veins, to *take away* metabolic waste products surrendered by the muscle into the blood. These blood vessels generally enter through the central part of the muscle, but can also enter towards one end. Thereafter, they branch into a capillary plexus, which spreads throughout the intermuscular septa, to eventually penetrate the endomysium around each muscle fibre. During exercise the capillaries dilate, increasing the amount of blood flow in the muscle by up to 800 times. The muscle tendon, because it is composed of a relatively inactive tissue, has a much less extensive blood supply.

Nerve Supply

The nerve supply to a muscle usually enters at the same place as the blood supply, and branches through the connective tissue septa into the endomysium in a similar way. Each skeletal muscle fibre is supplied by a single nerve ending. This is in contrast to other muscle tissues, which are able to contract without any nerve stimulation.

The nerve entering the muscle usually contains roughly equal proportions of sensory and motor nerve fibres, although some muscles may receive separate sensory branches. As the nerve fibre approaches the muscle fibre, it divides into a number of terminal branches, collectively called a *motor end plate*.

Motor Unit of a Skeletal Muscle
A motor unit consists of a single motor nerve cell and the muscle fibres stimulated by it. The motor units vary in size, ranging from cylinders of muscle 5–7mm in diameter in the upper limb and 7–10mm in diameter in the lower limb. The average number of muscle fibres within a unit is 150 (but this number ranges from less than 10 to several hundred). Where fine gradations of movement are required, as in the muscles of the eyeball or fingers, the number of muscle fibres supplied by a single nerve cell is small. On the other hand, where more gross movements are required, as in the muscles of the lower limb, each nerve cell may supply a motor unit of several hundred fibres.

The muscle fibres in a single motor unit are spread throughout the muscle, rather than being clustered together. This means that stimulation of a single motor unit will cause the entire muscle to exhibit a weak contraction.

Skeletal muscles work on an '*All or Nothing Principle*'. In other words, groups of muscle cells, or fasciculi, can either contract or not contract. Depending on the strength of contraction required, a certain number of muscle cells will contract totally, while others will not contract at all. When a great muscular effort is needed, most of the motor units may be stimulated at the same time. However, under normal conditions, the motor units tend to work in relays, so that during prolonged contractions some are resting while others are contracting.

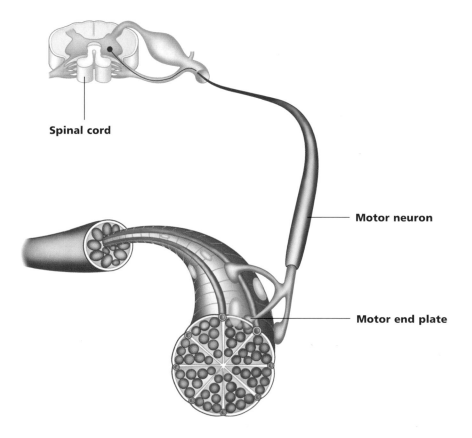

Spinal cord

Motor neuron

Motor end plate

Figure 7.5: A motor unit of a skeletal muscle.

Muscle Reflexes

Within skeletal muscles there are two specialized types of nerve receptors that can sense stretch. These are muscle spindles and Golgi tendon organs (GTO's). Muscle spindles are cigar-like in shape and consist of tiny modified muscle fibres called intrafusal fibres, and nerve endings, encased together within a connective tissue sheath. They lie between and parallel to the main muscle fibres. The GTO's are located mostly at the junction of muscles and their tendons or aponeurosis.

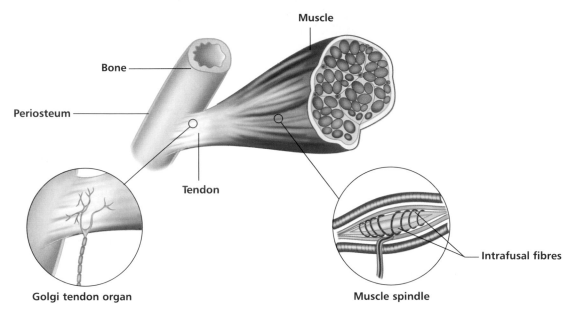

Figure 7.6: Anatomy of the muscle spindle and Golgi tendon organ.

Stretch Reflex

The stretch reflex helps control posture by maintaining muscle tone. It also helps prevent injury, by enabling a muscle to respond to a sudden or unexpected increase in length. The way it works is as follows:

1. When a muscle is lengthened, the muscle spindles are also stretched, causing each spindle to send a nerve impulse to the spinal cord.

2. On receiving this impulse, the spinal cord immediately sends an impulse back to the stretched muscle fibres, causing them to contract, in order to resist further stretching of the muscle. This circular process is known as a *reflex arc*.

3. An impulse is simultaneously sent from the spinal cord to the antagonist of the contracting muscle (i.e. the muscle opposing the contraction), causing the antagonist to relax, so that it cannot resist the contraction of the stretched muscle. This process is known as *reciprocal inhibition*.

4. Concurrent with this spinal reflex, nerve impulses are also sent up the spinal cord to the brain to relay information on muscle length and the speed of muscle contraction. A reflex in the brain feeds nerve impulses back to the muscle to ensure the appropriate muscle tone is maintained to meet the requirements of posture and movement.

5. Meanwhile, the stretch sensitivity of the minute intrafusal muscle fibres within the muscle spindle are smoothed and regulated by *gamma efferent nerve fibres*, arising from motor neurons within the spinal cord. Thus, a *gamma motor neuron reflex arc* ensures the evenness of muscle contraction, which would otherwise be jerky if muscle tone relied on the stretch reflex alone.

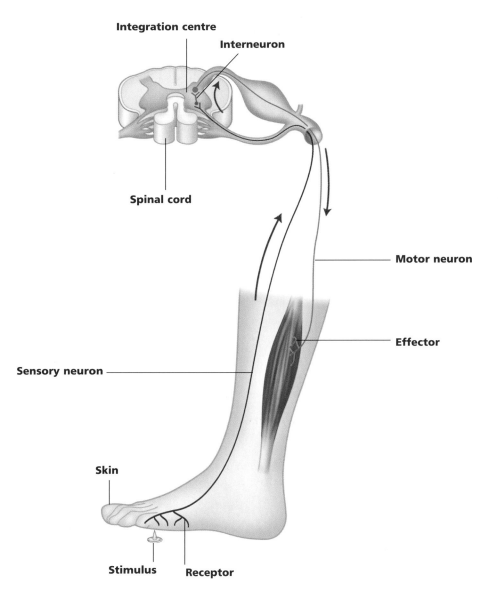

Figure 7.7: The basic reflex arc.

The classic clinical example of the stretch reflex in action is the knee jerk, or patellar reflex; whereby the patellar tendon is lightly struck with a small rubber hammer. This results in the following sequence of events:

1. The sudden stretch of the patellar tendon causes the quadriceps to be stretched.

2. This rapid stretch is registered by the muscle spindles within the quadriceps, causing the quadriceps to contract. This causes a small kick as the knee straightens suddenly, and takes the tension off the muscle spindles.

3. Simultaneously, nerve impulses to the hamstrings, which are the antagonists of the quadriceps, are inhibited, causing the hamstrings to relax.

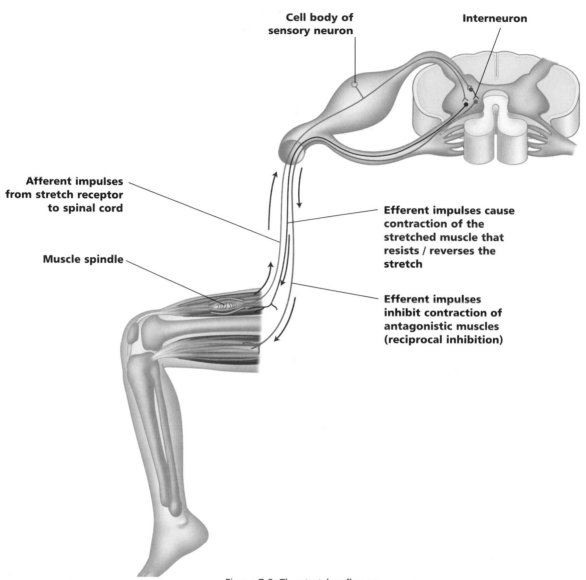

Figure 7.8: The stretch reflex arc.

Another obvious example of the stretch reflex in action is: when a person falls asleep in the sitting position, the head will relax forward, then jerk back up, because the stretched muscle spindles in the back of the neck have activated a reflex arc.

The stretch reflex also works constantly to maintain the tonus of our postural muscles. That is, it enables us to remain standing without conscious effort and without collapsing forwards. The sequence of events preventing this forward collapse occurs in a fraction of a second, as follows:

1. In standing, we naturally begin to sway forwards.

2. This pulls our calf muscles into a lengthened position, activating the stretch reflex.

3. The calf muscles consequently contract to pull us back to the upright position.

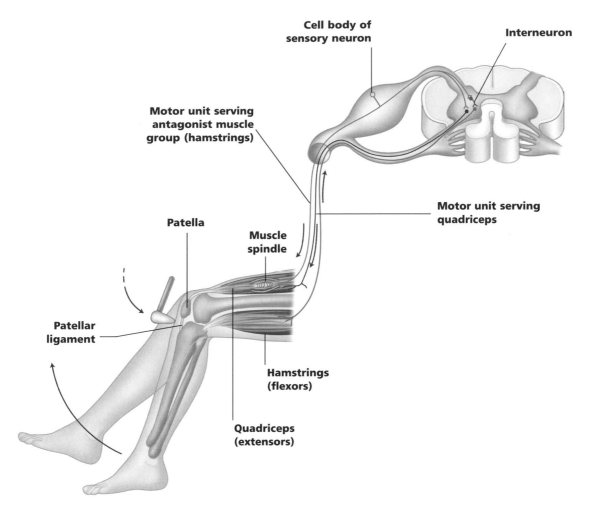

Figure 7.9: The patellar reflex.

Deep Tendon Reflex (Autogenic Inhibition)

In contrast to the stretch reflex, which involves the muscle spindle's response to muscle elongation, the deep tendon reflex involves the reaction of Golgi tendon organs (GTO's) to muscle contraction. As such, the deep tendon reflex creates the opposite effect to the stretch reflex. The way it works is as follows:

1. When a muscle contracts, it pulls on the tendons which are situated at either end of the muscle.

2. The tension in the tendon causes the GTO's to transmit impulses to the spinal cord, (some impulses continue on to the cerebellum).

3. As these impulses reach the spinal cord, they inhibit the motor nerves supplying the contracting muscle, causing it to relax.

4. Simultaneously, the motor nerves supplying the antagonist muscle are activated, causing it to contract. This process is called *reciprocal activation*.

5. Meanwhile, the information reaching the cerebellum is processed and fed back to help readjust muscle tension.

The deep tendon reflex has a protective function, preventing the muscle from contracting so hard that it rips its attachment off the bone. It is therefore especially important during activities such as running, which involve rapid switching between flexion and extension.

Note however, that in normal day-to-day movement, tension in the muscles is not sufficient to activate the GTO's deep tendon reflex. By contrast, the threshold of the muscle spindle stretch reflex is set much lower, because it must constantly maintain sufficient tonus in the postural muscles to keep us upright. Hence, it is active throughout normal daily activity.

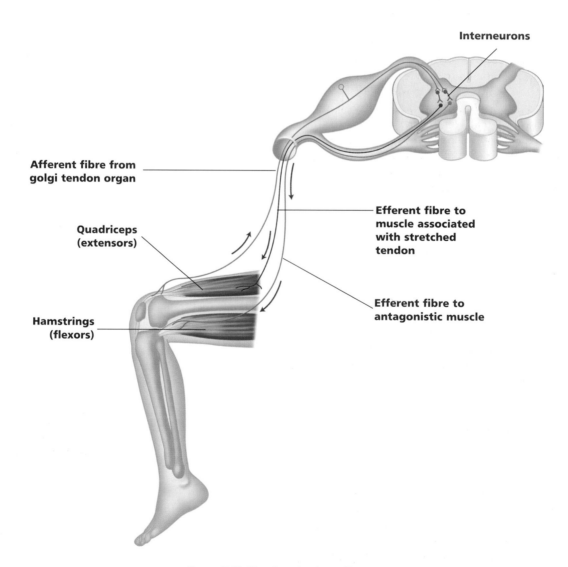

Interneurons

Afferent fibre from golgi tendon organ

Efferent fibre to muscle associated with stretched tendon

Quadriceps (extensors)

Efferent fibre to antagonistic muscle

Hamstrings (flexors)

Figure 7.10: The deep tendon reflex.

Isometric and Isotonic Contractions

A muscle will contract upon stimulation, in an attempt to bring its attachments closer together, but this does not necessarily result in a shortening of the muscle. If no movement results from contraction, such a contraction is called *isometric*. If the contraction of muscle results in the muscle creating movement of some sort, the contraction is called *isotonic*.

Isometric

An isometric contraction occurs when a muscle increases its tension, but the length of the muscle is not altered. In other words, although the muscle tenses, the joint over which the muscle works does not move. One example of this is holding a heavy object in the hand with the elbow held stationary and bent at 90 degrees. Trying to lift something that proves to be too heavy to move is another example. Note also that some of the postural muscles are largely working isometrically by automatic reflex. For example, in the upright position, the body has a natural tendency to fall forward at the ankle. This is prevented by isometric contraction of the calf muscles. Likewise, the centre of gravity of the skull would make the head tilt forwards if the muscles at the back of the neck did not contract isometrically to keep the head centralized.

Isotonic

It is the isotonic contractions of muscle that enable us to move about. Such contractions are of two types:

Concentric

In concentric contractions, the muscle attachments move closer together, causing movement at the joint. Using the example of holding an object in the hand, if the biceps muscle contracts concentrically, the elbow joint will flex and the hand will move towards the shoulder. Similarly, if we look up at the stars, the muscles at the back of the neck must contract concentrically to tilt the head back and extend the neck.

Eccentric

Eccentric contraction means that the muscle fibres 'pay out' in a controlled manner to slow down movements which gravity, if unchecked, would otherwise cause to be too rapid. For example, lowering an object held in the hand down to your side. Another example is simply sitting down into a chair. Therefore, the difference between concentric and eccentric contraction is that in the former, the muscle shortens, and in the latter, it actually lengthens.

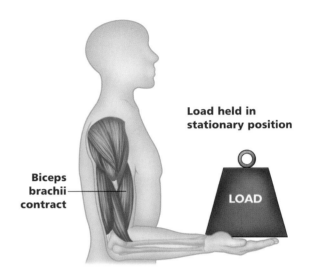

Load held in stationary position

Biceps brachii contract

LOAD

Figure 7.11: Isometric contraction.

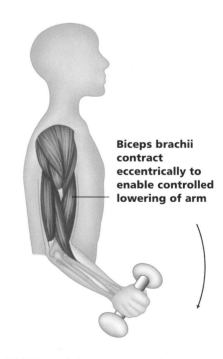

Biceps brachii contract eccentrically to enable controlled lowering of arm

Figure 7.12: Eccentric isotonic contraction.

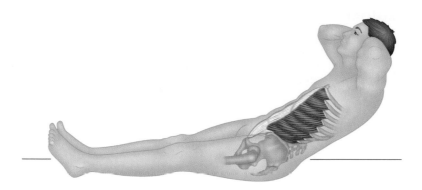

Figure 7.13: Abdominals contract to raise body concentrically.

Muscle Shape (Arrangement of Fascicles)

Muscles come in a variety of shapes according to the arrangement of their fascicles. The reason for this variation is to provide optimum mechanical efficiency for a muscle in relation to its position and action. The most common arrangement of fascicles give muscle shapes described as parallel, pennate, convergent and circular. Each of these shapes has further sub-categories.

Parallel

This arrangement has the fascicles running parallel to the long axis of the muscle. If the fascicles extend throughout the length of the muscle, it is known as a *strap muscle*, for example: sartorius (*see* figure 7.14). If the muscle also has an expanded belly and tendons at both ends, it is called a *fusiform* muscle, for example, the biceps brachii of the arm (*see* figure 7.14). A modification of this type of muscle has a fleshy belly at either end, with a tendon in the middle. Such muscles are referred to as *digastric*.

Pennate

Pennate muscles are so named because their short fasciculi are attached obliquely to the tendon, like the structure of a feather (penna = feather). If the tendon develops on one side of the muscle, it is referred to as *unipennate*, for example, the flexor digitorum longus in the leg (*see* figure 7.14). If the tendon is in the middle and fibres are attached obliquely from both sides, it is known as *bipennate*, of which the rectus femoris is a good example (*see* figure 7.14). If there are numerous tendinous intrusions into the muscle with fibres attaching obliquely from several directions, thus resembling many feathers side by side, the muscle is referred to as *multipennate*; the best example being the middle part of the deltoid muscle (*see* figure 7.14).

Convergent

Muscles that have a broad origin with fascicles converging toward a single tendon, giving the muscle a triangular shape, are called convergent muscles. The best example is the pectoralis major (*see* figure 7.14).

Circular

When the fascicles of a muscle are arranged in concentric rings, the muscle is referred to as *circular*. All the sphincter skeletal muscles in the body are of this type; i.e. they surround openings, which they close by contracting. An example includes the orbicularis oculi (*see* figure 7.14).

When a muscle contracts, it can shorten by up to 70% of its original length. Hence, the longer the fibres, the greater the range of movement. On the other hand, the strength of a muscle depends on the total number of muscle fibres it contains, rather than their length. Therefore:

1. Muscles with long parallel fibres produce the greatest range of movement, but are not usually very powerful.

2. Muscles with a pennate pattern, especially if multipennate, pack in the most fibres. Such muscles shorten less than long parallel muscles, but tend to be much more powerful.

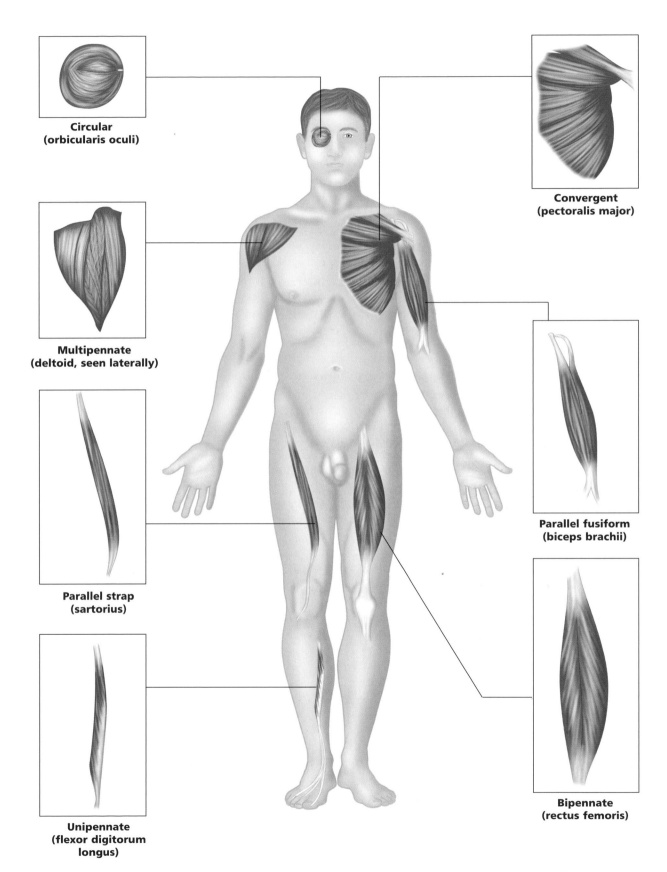

**Circular
(orbicularis oculi)**

**Convergent
(pectoralis major)**

**Multipennate
(deltoid, seen laterally)**

**Parallel fusiform
(biceps brachii)**

**Parallel strap
(sartorius)**

**Unipennate
(flexor digitorum
longus)**

**Bipennate
(rectus femoris)**

Figure 7.14: Muscle shapes.

Functional Characteristics of a Skeletal Muscle

All that has been said about muscles so far in this book enables us to formulate a list of functional characteristics pertaining to skeletal muscle.

Excitability

Excitability is the ability to receive and to respond to a stimulus. In the case of a muscle, when a nervous impulse from the brain reaches the muscle, a chemical known as *acetylcholine* is released. This chemical produces a change in the electrical balance in the muscle fibre and as a result generates an electrical current, known as an *action potential*. The action potential conducts the electrical current from one end of the muscle cell to the other and results in a contraction of the muscle cell, or muscle fibre (remember that one muscle cell = one muscle fibre).

Contractility

Contractility is the ability of a muscle to shorten forcibly when stimulated. In other words, the muscles themselves can only contract. They cannot lengthen, except via some external means (i.e. manually), beyond their normal resting length (*see Tonus* below). In other words, muscles can only pull their ends together (contract); they cannot push them apart.

Extensibility

Extensibility is the ability of a muscle to be extended, or returned to its resting length (which is a semi-contracted state), or slightly beyond. For example, if we bend forward at the hips from standing, the muscles of the back, such as erector spinae, lengthen eccentrically (*see page 127*) to lower the trunk, paying out slightly beyond their normal resting length, and are thus effectively 'elongated'.

Elasticity

Elasticity describes the ability of a muscle fibre to recoil after being lengthened, and therefore resume its resting length when relaxed. In a whole muscle, the elastic effect is supplemented by the important elastic properties of the connective tissue sheaths (endomysium and epimysium). Tendons also contribute some elastic properties. An example of this elastic recoil effect can be experienced when coming back up from a forward bend at the hips as described above, there is initially no muscle contraction. Instead, the upward movement is initiated purely by elastic recoil of the back muscles, after which, contraction of the back muscles completes the movement.

Tonus

Tonus, or muscle tone, is the term used to describe the slightly contracted state which muscles resume during the resting state. Muscle tonus does not produce active movements, but it keeps the muscles firm, healthy, and ready to respond to stimulation. It is the tonus of skeletal muscles that also helps stabilize and maintain posture. *Hypertonic* muscles are those muscles whose 'normal' resting state is over-contracted.

General Functions of Skeletal Muscles

Enable Movement

Skeletal muscles are responsible for all locomotion and manipulation, and they enable you to respond quickly.

Maintain Posture

Skeletal muscles support an upright posture against the pull of gravity.

Stabilize Joints

Skeletal muscles and their tendons stabilize joints.

Generate Heat

In common with smooth and cardiac muscles, skeletal muscles generate heat, which is important in maintaining a normal body temperature.

Musculo-skeletal Mechanics

Origins and Insertions

In the majority of movements, one attachment of a muscle remains relatively stationary while the attachment at the other end moves. The more stationary attachment is called the *origin* of the muscle, and the other attachment is called the *insertion*. A spring that closes a gate could be said to have its origin on the gatepost and its insertion on the gate itself. In the body, the arrangement is rarely so clear-cut, because depending on the activity one is engaged in, the fixed and moveable ends of the muscle may be reversed. For example, muscles that attach the upper limb to the chest normally move the arm relative to the trunk; which means their origins are on the trunk and their insertions are on the upper limb. However, in climbing, the arms are fixed, while the trunk is moved as it is pulled up to the fixed limbs. In this type of situation, where the insertion is fixed and the origin moves, the muscle is said to perform a *reversed action*. Because there are so many situations where muscles are working with a reversed action, it is sometimes less confusing to simply speak of 'attachments', without reference to origin and insertion.

In practice, muscle attachments that lay more proximally, i.e. more towards the trunk or on the trunk, are usually referred to as the origin. Attachments that lie more distally, i.e. away from the attached end of a limb, or away from the trunk, are referred to as the insertion.

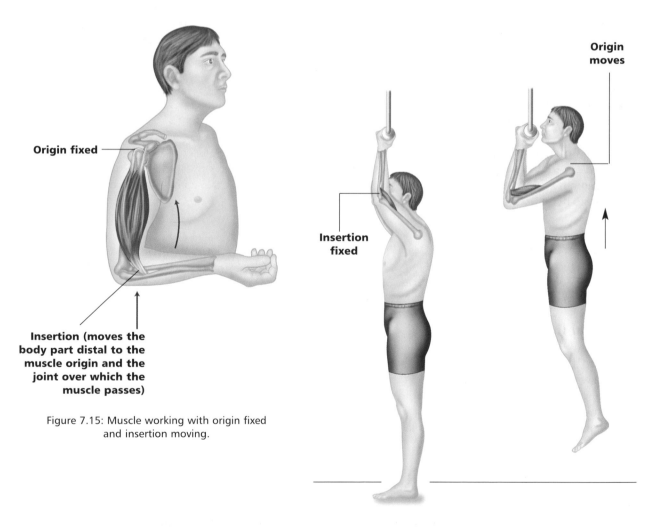

Origin fixed

Insertion (moves the body part distal to the muscle origin and the joint over which the muscle passes)

Figure 7.15: Muscle working with origin fixed and insertion moving.

Origin moves

Insertion fixed

Figure 7.16: Climbing: muscles are working with insertion fixed and origin moving (reversed action).

Group Action of Muscles

Muscles work together, or in opposition, to achieve a wide variety of movements. Therefore, whatever one muscle can do, there is another muscle that can undo it. Muscles may also be required to provide additional support or stability to enable certain movements to occur elsewhere.

Muscles are classified into four functional groups:

• Prime Mover or Agonist
• Antagonist
• Synergist
• Fixator

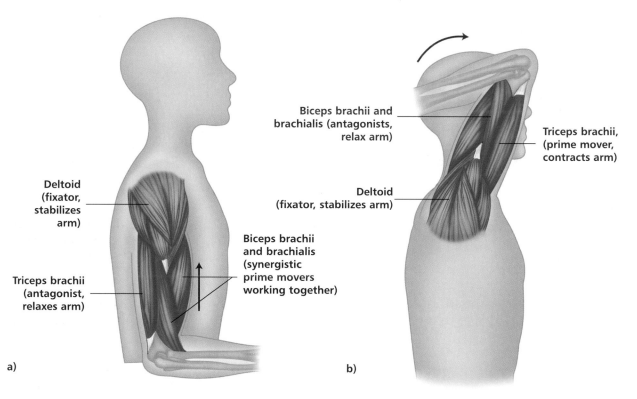

Figure 7.17: Group action of muscles; a) flexing arm at elbow,
b) extending arm at elbow (showing reversed roles of prime mover and antagonist).

Prime Mover or Agonist

A prime mover (also called an agonist) is a muscle that contracts to produce a specified movement. An example is the biceps brachii, which is the prime mover of elbow flexion. Other muscles may assist the prime mover in providing the same movement, albeit with less effect. Such muscles are called assistant or secondary movers. For example, the brachialis assists the biceps brachii in flexing the elbow, and is therefore a secondary mover.

Antagonist

The muscle on the opposite side of a joint to the prime mover, and which must relax to allow the prime mover to contract, is called an antagonist. For example, when the biceps brachii on the front of the arm contract to flex the elbow, the triceps brachii on the back of the arm must relax to allow this movement to occur. When the movement is reversed, i.e. when the elbow is extended, the triceps brachii becomes the prime mover and the biceps brachii assumes the role of antagonist.

Synergist

Synergists prevent any unwanted movements that might occur as the prime mover contracts. This is especially important where a prime mover crosses two joints, because when it contracts it will cause movement at both joints, unless other muscles act to stabilize one of the joints. For example, the muscles that flex the fingers not only cross the finger joints, but also cross the wrist joint, potentially causing movement at both joints. However, it is because you have other muscles acting synergistically to stabilize the wrist joint that you are able to flex the fingers into a fist without also flexing the wrist at the same time.

A prime mover may have more than one action, so synergists also act to eliminate the unwanted movements. For example, the biceps brachii will flex the elbow, but its line of pull will also supinate the forearm (twist the forearm, as in tightening a screw). If you want flexion to occur without supination, other muscles must contract to prevent this supination. In this context, such synergists are sometimes called neutralizers.

Fixator

A synergist is more specifically referred to as a fixator or stabilizer when it immobilizes the bone of the prime mover's origin, thus providing a stable base for the action of the prime mover. The muscles that stabilize (fix) the scapula during movements of the upper limb are good examples. The sit-up exercise gives another good example: The abdominal muscles attach to both the ribcage and the pelvis. When they contract to enable you to perform a sit-up, the hip flexors will contract synergistically as fixators to prevent the abdominals tilting the pelvis; enabling the upper body to curl forward as the pelvis remains stationary.

Leverage

The bones, joints, and muscles together form a system of levers in the body, in order to optimize the relative strength, range and speed required of any given movement. The joints act as the fulcra (sing. *fulcrum*), while the muscles apply the effort and the bones bear the weight of the body part to be moved.

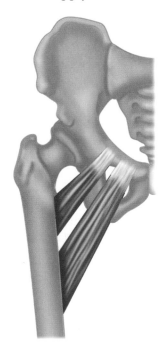

A muscle attached close to the fulcrum will be relatively weaker than it would be if it were attached further away. However, it is able to produce a greater range and speed of movement; because the length of the lever amplifies the distance travelled by its moveable attachment. Figure 7.18 illustrates this in relation to the adductors of the hip joint. The muscle so positioned to move the greater load (in this case, adductor longus) is said to have a *mechanical advantage*. The muscle attached close to the fulcra is said to operate at a *mechanical disadvantage*, although it can move a load more rapidly through larger distances.

Figure 7.18: The pectineus is attached closer to the axis of movement than the adductor longus. Therefore, the pectineus is the weaker adductor of the hip, but is able to produce a greater movement of the lower limb per centimetre of contraction.

The following illustrations depict the differences in first, second and third class levers, with examples in the human body.

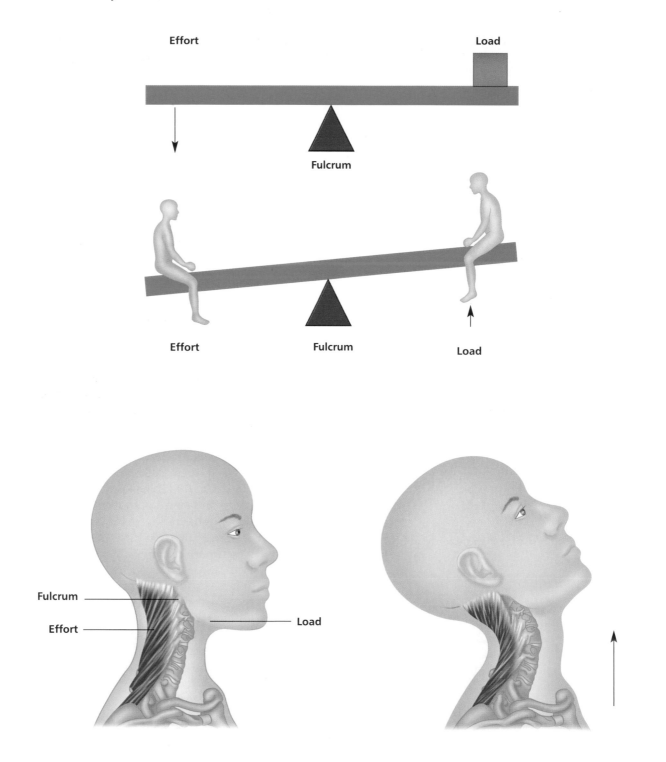

Figure 7.19: First-class lever: The relative position of components is Load-Fulcrum-Effort. Examples include a seesaw (as above). Another example is a pair of scissors. In the body, an example is the ability to extend the head and neck, i.e. the facial structures are the load; the atlanto-occipital joint is the fulcrum; the posterior neck muscles provide the effort.

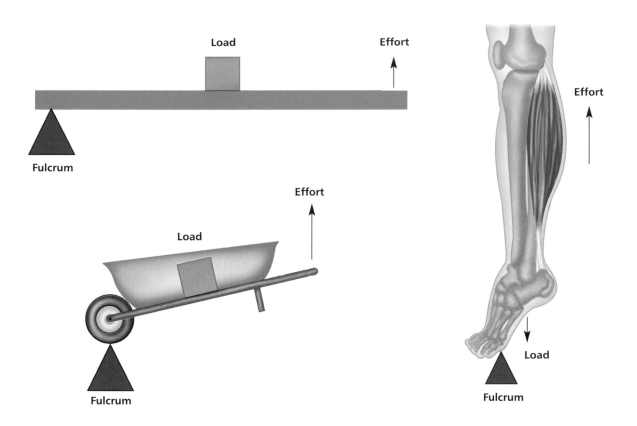

Figure 7.20: Second-class lever: The relative position of components is Fulcrum-Load-Effort. The best example is a wheelbarrow. In the body, an example is the ability to raise the heels off the ground in standing, i.e. the ball of the foot is the fulcrum; the body-weight is the load; the calf muscles provide the effort. With second-class levers, speed and range of movement are sacrificed for strength.

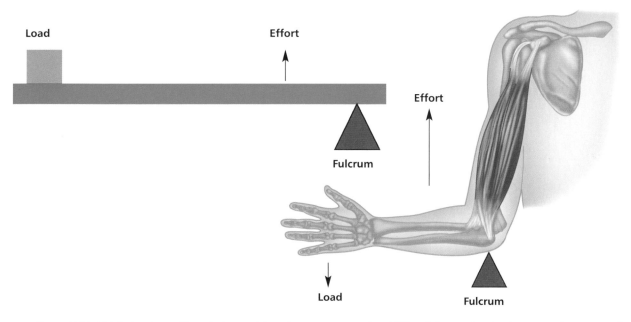

Figure 7.21: Third-class lever: The relative position of components is Load-Effort-Fulcrum. A pair of tweezers is an example of this. In the body, most skeletal muscles act in this way. An example is flexing the forearm, i.e. an object held in the hand is the load; the biceps provide the effort; the elbow joint is the fulcrum. With third-class levers, strength is sacrificed for speed and range of movement.

Factors in Muscles That Limit Skeletal Movement

The inability of a muscle to contract or lengthen beyond a certain point can cause some practical hindrances to bodily movement; which are outlined as follows:

Passive Insufficiency

Muscles that span over two joints are called biarticular muscles. These muscles may be unable to 'pay out' sufficiently to allow full movement of both joints simultaneously, unless the muscle has been trained to relax. For example, most people need to bend their knees in order to touch their toes. This is because the hamstrings (which span the hip and knee joints) cannot lengthen enough to allow full flexion at the hip joint without also pulling the knee joint into flexion. For the same reason, it is easier to pull your thigh to your chest if your knee is bent than it is with your knee straight. This limitation is called *passive insufficiency*. Passive insufficiency is therefore the inability of a muscle to lengthen by more than a fixed percentage of its length.

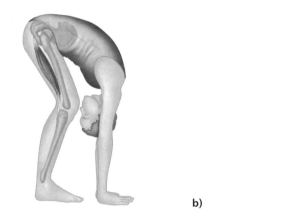

a)   b)

Figure 7.22: Passive insufficiency example 1; a) having to bend the knees to touch the toes means there is passive insufficiency of the hamstrings, and b) being able to touch the toes with the knees straight means there is much less passive insufficiency of the hamstrings.

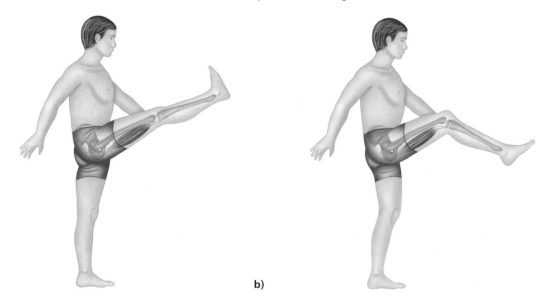

a)   b)

Figure 7.23: Passive insufficiency example 2; a) a high kick with knee straight is possible only if the hamstrings have been trained to overcome their passive insufficiency, and b) for most people, an attempt at a high kick will be restricted by hamstring passive insufficiency causing the knee to bend.

Active Insufficiency

Active insufficiency is the opposite of passive insufficiency. Whereas passive insufficiency results from the inability of a muscle to *lengthen* by more than a fixed percentage of its length, active insufficiency results from the inability of a muscle to *contract* by more than a fixed amount. For example, most people can flex their knee to bring their heel close to their buttock, if their hip is flexed; because the upper part of the hamstrings are lengthened and the lower part is shortened. However, one is normally unable to fully flex the knee when the hip is extended. This is because with hip extended, the hamstrings are already shortened, meaning that there is insufficient 'shortening' potential remaining in the hamstrings to then fully flex the knee.

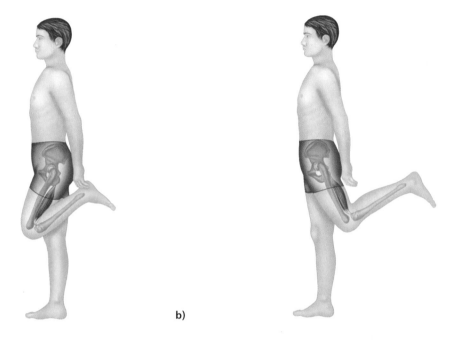

a)   b)

Figure 7.24: Active insufficiency; a) with hip flexed, the hamstrings are stretched at the hip, enabling its contraction to fully flex the knee, and b) with hip extended, the shortened hamstrings are unable to contract still further to fully flex the knee.

Concurrent Movement

If extension of the hip is required at the same time as extension of the knee, as in the push off from the ground in running, the phenomenon known as *concurrent movement* applies, and proves very useful. To grasp the concept of concurrent movement, first remember that when the hamstrings contract, they are able to both extend the hip joint and flex the knee joint, either singly or simultaneously. So, in analyzing the example of running in more detail, we observe the following:

• As the foot pushes against the ground, the hamstrings contract to extend the hip.
• Meanwhile, fixators prevent the hamstrings from flexing the knee.
• Therefore, the hamstrings are shortened only at their upper end (origin), but remain lengthened at their lower end (insertion).
• The antagonist to the hamstring's action of flexing the hip is the rectus femoris, which relaxes because of reciprocal inhibition (*see* page 122) to allow the hamstrings to contract.
• When the hip is well extended, the already stretched rectus femoris is unable to lengthen further, causing it to pull the knee into extension.
• Therefore, the rectus femoris is lengthened at its upper end and shortened at its lower end.

Concurrent movement therefore avoids passive and active insufficiency of the hamstrings and rectus femoris by neither shortening nor stretching both ends of either muscle, but rather, having one end lengthen as the other shortens, and vice versa in the other muscle. Figure 7.25 should hopefully elucidate this concept.

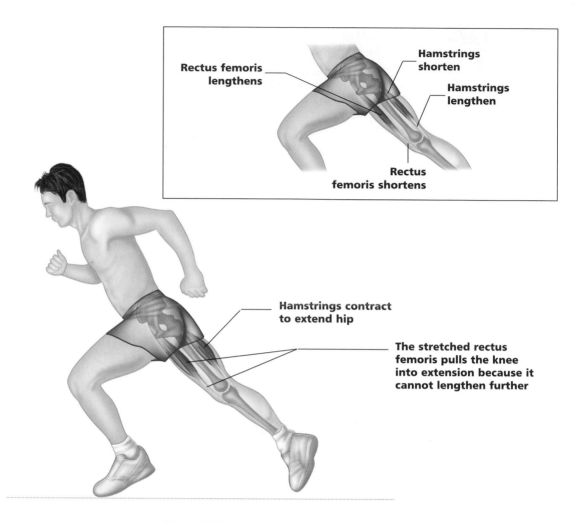

Figure 7.25: Concurrent movement.

Countercurrent Movement

If flexion of the hip is required to occur at the same time as extending the knee, as in kicking a ball, a countercurrent movement occurs. So, in analyzing the example of kicking in more detail, we observe the following:

• To kick a ball, the rectus femoris acts as a prime mover to flex the hip and extend the knee.
• Thus, both the upper and lower portions of rectus femoris are shortened.
• The hamstrings relax due to reciprocal inhibition, so that they can extend at both ends and allow the kick to occur.
• The rectus femoris relaxes once the movement has been made, but the momentum of the movement is still propelling the leg forward.
• At this stage, the hamstrings contract to act as a 'brake' for the leg, as it flies forward.

Countercurrent movements therefore prevent injury by ensuring the antagonist relaxes first, then contracts at the right time to prevent the forces of momentum from overstretching muscles and ligaments. So called ballistic movements are relying on this principle, but are often done so forcefully that the power of momentum is greater than the ability of the antagonist to 'brake' that momentum. In such instances, muscle and ligament damage often occurs.

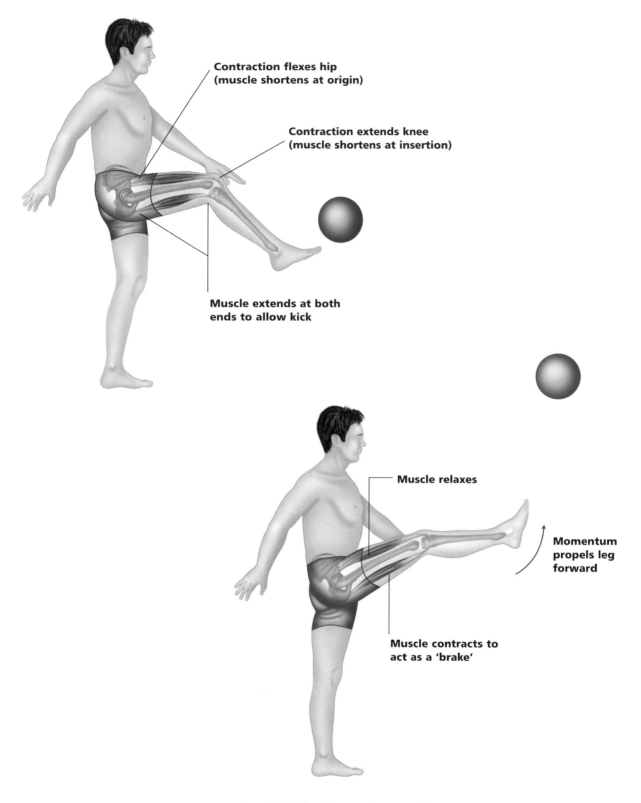

Contraction flexes hip
(muscle shortens at origin)

Contraction extends knee
(muscle shortens at insertion)

Muscle extends at both
ends to allow kick

Muscle relaxes

Momentum
propels leg
forward

Muscle contracts to
act as a 'brake'

Figure 7.26: Countercurrent movement.

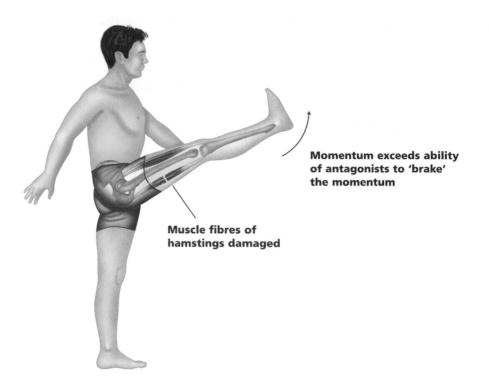

Momentum exceeds ability
of antagonists to 'brake'
the momentum

Muscle fibres of
hamstings damaged

Figure 7.27: Damage that can be caused by an over zealous ballistic stretch.

Core Stability

During day-to-day activity, skeletal muscles are acting as either stabilizing muscles or muscles of movement (as outlined under Group Action of Muscles, *see* page 132). Stabilizing muscles maintain posture or hold the body in a given position as a 'platform' so that other muscles can cause the body to move in some way.

Stabilizing muscles tend to be situated deep within the body. To maintain posture or a steady 'platform', their fibres perform a minimal contraction over an extended period of time. Hence they are built for endurance and therefore have many slow-twitch fibres (*see* 'red and white muscle fibres', *see* page 120). Persons with poor postural alignment or an inactive lifestyle tend to have insufficient tone in these muscles, which further exacerbates their poor posture and lessens their ability to stabilize functional movements.

When the stabilizing muscles are under-used, the nerve impulses find it more difficult to get through to those muscles, leading to what is referred to as *poor recruitment*. That means, if we do not use a muscle for an extended period of time, we will find it more difficult to re-enervate that muscle back into use. Consequently, the majority of people in modern society would benefit from exercises that specifically address their neglected deep postural muscles.

It is particularly important to maintain your torso as a stable platform relative to the movements carried out by your limbs. As your torso or mid-section is the 'core' of your body, its success as a stable platform is referred to as *core stability*. Good core stability therefore allows you to maintain a rigid mid-section without gravity or other forces interfering with the movement you wish to perform. Core stability muscles can be retrained, especially through bracing and stabilizing exercises; a fact utilized in physiotherapy treatment, Pilates, Taiji Quan, Hatha Yoga and so on. In essence, core stability can be summarized as the successful recruitment of deep muscles that maintain the natural curvatures (neutral alignment) of the spine during all other movements of the body.

Good core stability results from the deep stabilizing trunk muscles co-ordinating their contraction to stabilize the spine, rather like the tightening of guide ropes around a pole or mast to give it strength and maintain its position.

The deep stabilizing or 'core stability' muscles collectively create what is known as an 'inner unit' of muscle. These muscles include the tranversus abdominis, multifidis, pelvic floor, diaphragm and posterior fibres of the internal oblique. The main muscles that initiate movement of the limbs whilst working in unison with the inner unit are collectively referred to as the 'outer unit' or global muscles; i.e. the spinal erectors, external and internal obliques, latissimus dorsi, gluteals, hamstrings and adductors. The following effects summarize how core stability is enhanced by some subsidiary factors of body mechanics:

Thoraco-lumbar fascia gain
As the abdominal wall is pulled in by the contraction of the transversus abdominis, the internal oblique acts synergistically to pull upon the thoraco-lumbar fascia (which wraps around the spine, connecting the deep trunk muscles to it). This in turn exerts a force on the lumbar spine that helps support and stabilize it (this force is called thoraco-lumbar gain). More specifically, the increased tension of the thoraco-lumbar fascia compresses the erector spinae and multifidis muscles, encouraging these to contract and resist the forces that are trying to flex the spine. The classic analogy is that of the guide ropes of a tent acting together to support the main structure of the tent.

Research demonstrates that in addition to the above, the paraspinal muscles – interspinalis and intertransversarii – assist core stability insofar as they provide an individual stabilizing effect on their adjacent vertebrae, acting in a similar way to ligaments.

It is not just the recruitment of these deep-trunk muscles that is significant, but also how and when they are recruited that is important. Two key researchers in core stability theory, Hodges and Richardson, showed that co-contraction of the transversus abdominis and multifidis muscles occurs prior to any movement of the limbs. This suggests that these muscles anticipate dynamic forces that may act on the lumbar spine and stabilize the area before any movement takes place elsewhere.

Intra-abdominal pressure
Pressure in the abdominal cavity is increased as a result of the abdominal wall being pulled inwards by the transversus abdominis, along with a co-contraction of the pelvic floor, internal oblique and low back muscles. This in turn exerts a tensile force on the rectus sheath, which encloses the rectus abdominis muscle. Because the rectus sheath attaches to the internal oblique and transversus abdominis muscles, it effectively surrounds the abdomen. The tension of the rectus sheath therefore increases the pressure within the abdomen like a pressurized balloon. This further facilitates the stability of the core. In practice we clearly experience this when we hold our breath during a significant lifting or throwing action, during which time we can feel ourselves contract the diaphragm and pelvic floor muscles.

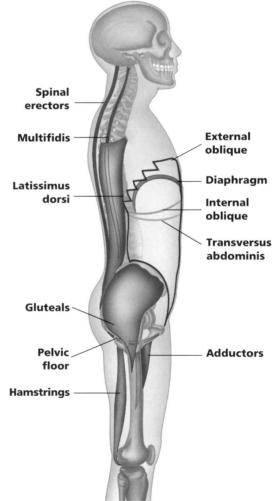

Figure 7.28: Schematic diagram of the core stability (inner unit) muscles and the global (outer unit) muscles.

Superficial and Deep Muscles of the Head and Neck (Lateral View)

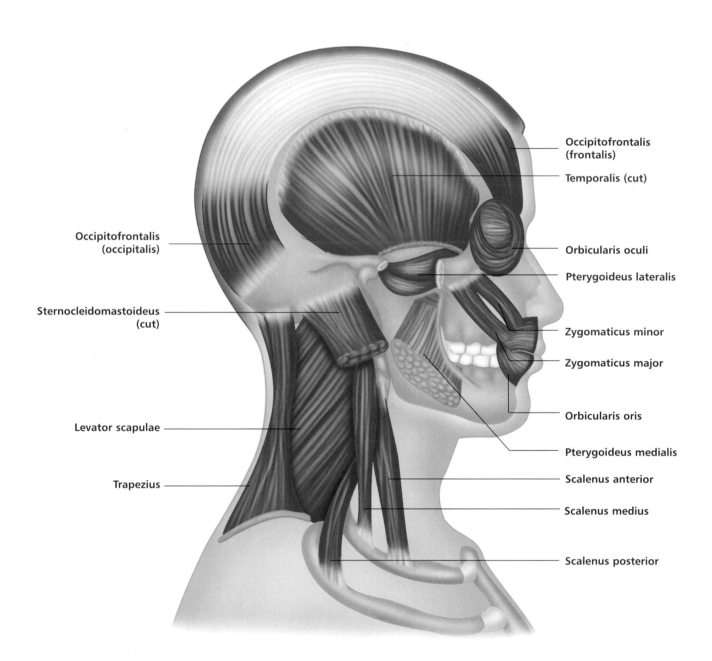

Occipitofrontalis (frontalis)

Temporalis (cut)

Orbicularis oculi

Pterygoideus lateralis

Zygomaticus minor

Zygomaticus major

Orbicularis oris

Pterygoideus medialis

Scalenus anterior

Scalenus medius

Scalenus posterior

Occipitofrontalis (occipitalis)

Sternocleidomastoideus (cut)

Levator scapulae

Trapezius

Muscle	Origin	Insertion	Action	Nerve
Occipitalis	Occipital bone. Mastoid process of temporal bone.	Galea aponeurotica.	Pulls scalp backward.	Facial **V11** nerve.
Frontalis	Galea aponeurotica.	Fascia and skin above eyes and nose.	Pulls scalp forwards.	Facial **V11** nerve.
Sternocleidomastoideus	Sternal head: anterior surface of upper sternum. Clavicular head: medial third of clavicle.	Mastoid process of temporal bone. Lateral third of superior nuchal line of occipital bone.	Contraction of both sides: flexes neck and draws head forward. Raises sternum, and consequently the ribs, during deep inhalation. Contraction of one side: tilts the head towards the same side. Rotates head to face the opposite side.	Accessory **X1** nerve; with sensory supply for proprioception from cervical nerves, C2–C3.
Temporalis	Temporal fossa, including parietal, temporal and frontal bones.	Coronoid process of mandible. Anterior border of ramus of mandible.	Closes jaw. Clenches teeth. Assists in side to side movement of mandible.	Anterior and posterior deep temporal nerves from the trigeminal **V** nerve (mandibular division).
Orbicularis oculi *Orbital part*	Frontal bone. Medial wall of orbit (on maxilla).	Circular path around orbit, returning to origin.	Strongly closes eyelids (firmly 'screws up' the eye).	Facial **V11** nerve (temporal and zygomatic branches).
Palpebral part	Medial palpebral ligament.	Lateral palpebral ligament into zygomatic bone.	Gently closes eyelids (and comes into action involuntarily, as in blinking).	Facial **V11** nerve (temporal and zygomatic branches).
Lacrimal part	Lacrimal bone.	Lateral palpebral raphe.	Dilates lacrimal sac and brings lacrimal canals onto surface of eye.	Facial **V11** nerve (temporal and zygomatic branches).
Pterygoideus lateralis	Superior head: lateral surface of greater wing of sphenoid. Inferior head: lateral surface of lateral pterygoid plate of sphenoid.	Superior head: capsule and articular disc of the temporomandibular joint. Inferior head: neck of mandible.	Protrudes mandible. Opens mouth. Moves mandible from side to side (as in chewing).	Trigeminal **V** nerve (mandibular division).
Pterygoideus medialis	Medial surface of lateral pterygoid plate of the sphenoid bone. Pyramidal process of the palatine bone. Tuberosity of maxilla.	Medial surface of the ramus and the angle of the mandible.	Elevates and protrudes the mandible. Therefore it closes the jaw and assists in side to side movement of the mandible, as in chewing.	Trigeminal **V** nerve (mandibular division).
Zygomaticus minor	Lower surface of zygomatic bone.	Lateral part of upper lip lateral to levator labii superioris.	Elevates the upper lip. Forms nasolabial furrow.	Facial **V11** nerve (buccal branches).
Zygomaticus major	Upper lateral surface of zygomatic bone.	Skin at corner of mouth. Orbicularis oris.	Pulls corner of mouth up and back, as in smiling.	Facial **V11** nerve (zygomatic and buccal branches).
Orbicularis oris	Muscle fibres surrounding the opening of mouth, attached to the skin, muscle and fascia of the lips and surrounding area.	Skin and fascia at corner of mouth.	Closes lips, compresses lips against teeth, protrudes (purses) lips, and shapes lips during speech.	Facial **V11** nerve (buccal and mandibular branches).
Scalenus (anterior, medius, posterior)	Transverse processes of cervical vertebrae, C2–C7.	Scalenus Anterior and Medius: First rib. Scalenus Posterior: Second rib.	Acting together: flex neck. Raise first rib during active respiratory inhalation. Individually: laterally flex and rotate neck.	Ventral rami of cervical nerves, C3–C8.

Superficial and Intermediate Muscles of the Upper Body (Anterior View)

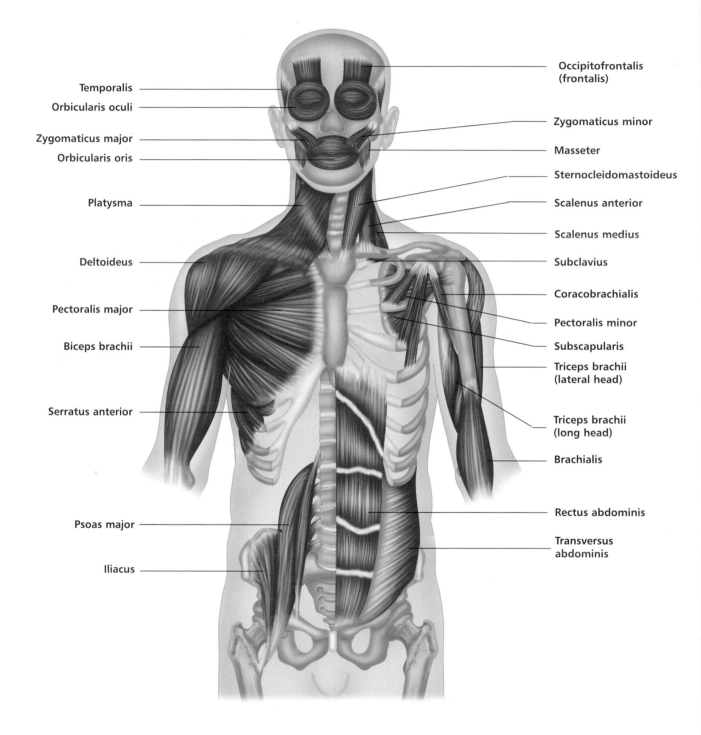

Temporalis

Orbicularis oculi

Zygomaticus major

Orbicularis oris

Platysma

Deltoideus

Pectoralis major

Biceps brachii

Serratus anterior

Psoas major

Iliacus

Occipitofrontalis (frontalis)

Zygomaticus minor

Masseter

Sternocleidomastoideus

Scalenus anterior

Scalenus medius

Subclavius

Coracobrachialis

Pectoralis minor

Subscapularis

Triceps brachii (lateral head)

Triceps brachii (long head)

Brachialis

Rectus abdominis

Transversus abdominis

Muscle	Origin	Insertion	Action	Nerve
Platysma	Subcutaneous fascia of upper quarter of chest.	Subcutaneous fascia and muscles of chin and jaw. Inferior border of mandible.	Pulls lower lip from corner of mouth downwards and laterally. Draws skin of chest upwards.	Facial **V11** nerve (cervical branch).
Deltoideus	Clavicle, acromion process, and spine of the scapula.	Deltoid tuberosity, situated half way down the lateral surface of the shaft of the humerus.	Anterior fibres: flex and medially rotate the humerus. Middle fibres: abduct the humerus at the shoulder joint. Posterior fibres: extend and laterally rotate the humerus.	Axillary nerve, C**5**, **6**, from the posterior cord of the brachial plexus.
Pectoralis major	Clavicular head: medial half or two thirds of front of clavicle. Sternocostal portion: sternum, and upper six costal cartilages. Rectus sheath.	Upper shaft of humerus.	Adducts and medially rotates the humerus. Clavicular portion: flexes and medially rotates the shoulder joint, and horizontally adducts humerus towards the opposite shoulder. Sternocostal portion: obliquely adducts humerus towards the opposite hip.	To upper fibres: lateral pectoral nerve, C**5**, **6**, **7**. To lower fibres: lateral and medial pectoral nerves, C**6**, **7**, **8**, **T1**.
Serratus anterior	Outer surfaces and superior borders of upper eight or nine ribs, and the fascia covering their intercostal spaces.	Anterior (costal) surface of the medial border of scapula and inferior angle of scapula.	Rotates scapula for abduction and flexion of arm. Protracts scapula.	Long thoracic nerve, C**5**, **6**, **7**, 8.
Masseter	Zygomatic process of maxilla. Medial and inferior surfaces of zygomatic arch.	Angle of ramus of mandible. Coronoid process of mandible.	Closes jaw. Clenches teeth. Assists in side to side movement of mandible.	Trigeminal **V** nerve (mandibular division).
Subclavius	Junction of the first rib and the first costal cartilage.	Floor of a groove on the lower (inferior) surface of the clavicle.	Depresses clavicle and draws it towards the sternum.	Nerve to subclavius, C**5**, **6**.
Coracobrachialis	Tip of the corocoid process of scapula.	Medial aspect of humerus at mid-shaft.	Weakly adducts shoulder joint. Helps stabilize humerus.	Musculocutaneous nerve, C**6**, **7**.
Pectoralis minor	Outer surfaces of third, fourth and fifth ribs and fascia of the corresponding intercostal spaces.	Corocoid process of scapula.	Draws scapula forward and downward. Raises ribs during forced inspiration.	Medial pectoral nerve with fibres from a communicating branch of the lateral pectoral nerve, C(6), **7**, **8**, **T1**.
Subscapularis	Subscapular fossa (anterior surface of scapula).	Lesser tubercle of humerus. Capsule of shoulder joint.	Stabilizes shoulder joint; mainly preventing the head of the humerus being pulled upwards by the deltoid, biceps and long head of triceps. Medially rotates humerus.	Upper and lower subscapular nerves, C**5**, **6**, 7, from the posterior cord of the brachial plexus.
Iliopsoas (Psoas major, Iliacus)	Psoas major: Transverse processes of lumbar vertebrae, (L1–L5). Bodies of thoracic vertebra and lumbar vertebrae, (T12–L5). Intervertebral discs above each lumbar vertebra. Iliacus: Superior two-thirds of iliac fossa. Anterior ligaments of the lumbosacral and sacroiliac joints.	Lesser trochanter of femur.	Main flexor of hip joint. Acting from its insertion, flexes the trunk, as in sitting up from the supine position.	Psoas major: Ventral rami of lumbar nerves, L1, **2**, **3**, 4. Iliacus: Femoral nerve, L(1), **2**, **3**, 4.
Rectus abdominis	Pubic crest and symphysis pubis.	Anterior surface of xiphoid process. Fifth, sixth and seventh costal cartilages.	Flexes lumbar spine. Depresses ribcage. Stabilizes the pelvis during walking.	Ventral rami of thoracic nerves, T5–12.

Superficial and Intermediate Muscles of the Upper Body (Posterior View)

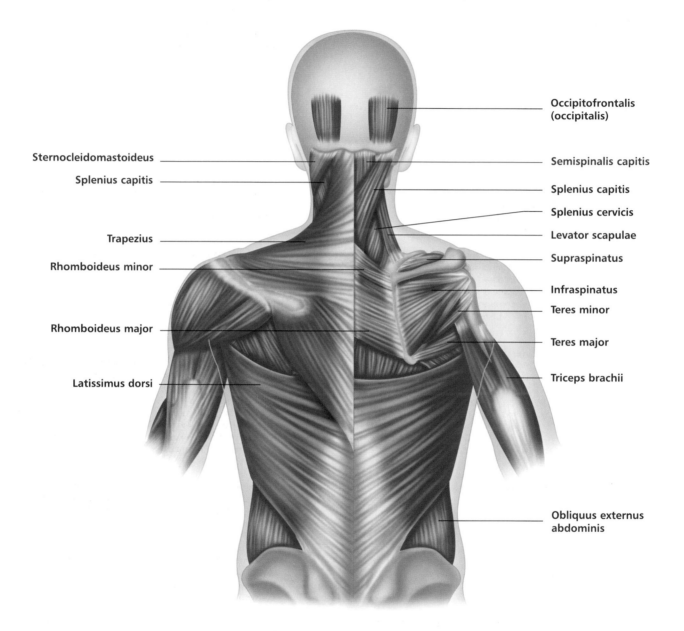

Occipitofrontalis
(occipitalis)

Sternocleidomastoideus

Splenius capitis

Semispinalis capitis

Splenius capitis

Splenius cervicis

Trapezius

Levator scapulae

Rhomboideus minor

Supraspinatus

Infraspinatus

Teres minor

Rhomboideus major

Teres major

Latissimus dorsi

Triceps brachii

Obliquus externus
abdominis

Muscle	Origin	Insertion	Action	Nerve
Trapezius	Occipital bone. Spinous processes of cervical vertebra, (C7) and thoracic vertebrae, (T1–T12).	Lateral third of clavicle. Acromion process. Spine of scapula.	Upper fibres: pull the shoulder girdle up. Helps prevent depression of the shoulder girdle. Middle fibres: retract (adduct) scapula. Lower fibres: depress scapula, particularly against resistance. Upper and lower fibres together: rotate scapula.	Accessory **X1** nerve. Ventral ramus of cervical nerves, C2, **3**, **4**.
Rhomboideus (minor, major)	Spinous processes of the cervical and thoracic vertebrae, (C7–T1).	Medial (vertebral) border of scapula.	Retracts (adducts) scapula. Stabilizes scapula. Slightly assists in outer range of adduction of arm.	Dorsal scapular nerve, C**4**, **5**.
Latissimus dorsi	Thoracolumbar fascia, which is attached to the spinous processes of vertebrae, (T7–S5). Posterior part of iliac crest. Lower three or four ribs. Inferior angle of the scapula.	Floor of the intertubercular sulcus (bicipital groove) of humerus.	Extends the flexed arm. Adducts and medially rotates the humerus. Assists in forced inspiration, by raising the lower ribs.	Thoracodorsal nerve, C**6**, **7**, **8**, from the posterior cord of the brachial plexus.
Splenius capitis / cervicis	Capitis: Spinous processes of the cervical vertebra, (C7) and thoracic vertebrae, (T1–T4). Cervicis: Spinous processes of thoracic vertebrae, (T3–T6).	Capitis: Mastoid process of temporal bone. Lateral part of superior nuchal line, deep to the attachment of the sternocleidomastoid. Cervicis: Transverse processes of cervical vertebrae, (C1–C3).	Acting together: extend the head and neck. Individually: laterally flexes neck. Rotates the face to the same side as contracting muscle.	Dorsal rami of middle and lower cervical nerves.
Levator scapulae	Transverse processes of cervical vertebrae, (C1–C4).	Upper medial (vertebral) border of the scapula.	Elevates scapula. Helps retract scapula. Helps bend neck laterally.	Dorsal scapular nerve, C**4**, **5** and cervical nerves, C**3**, **4**.
Supraspinatus	Supraspinous fossa of scapula.	Greater tubercle of the humerus. Capsule of shoulder joint.	Initiates the process of abduction at the shoulder joint, so that the deltoid can take over at the later stages of abduction.	Suprascapular nerve, C4, **5**, 6, from the upper trunk of the brachial plexus.
Infraspinatus	Infraspinous fossa of the scapula.	Greater tubercle of humerus. Capsule of shoulder joint.	Helps prevent posterior dislocation of the shoulder joint. Laterally rotates humerus.	Suprascapular nerve, C(4), **5**, 6, from the upper trunk of the brachial plexus.
Teres minor	Upper two-thirds of the lateral border of the dorsal surface of scapula.	Greater tubercle of humerus. Capsule of shoulder joint.	Helps prevent upward dislocation of the shoulder joint. Laterally rotates humerus. Weakly adducts humerus.	Axillary nerve, C**5**, **6**, from the posterior cord of the brachial plexus.
Teres major	Lower third of the posterior surface of the lateral border of the scapula.	Medial lip of the intertubercular sulcus (bicipital groove) of humerus.	Adducts humerus. Medially rotates humerus. Extends humerus from the flexed position.	Lower subscapular nerve, C5, **6**, 7, from the posterior cord of the brachial plexus.
Obliquus externus abdominis	Lower eight ribs.	Anterior half of iliac crest, and into an abdominal aponeurosis that terminates in the linea alba.	Compresses abdomen. Contraction of one side alone bends the trunk laterally to that side and rotates it to the opposite side.	Thoracic nerves, T5–T12.

Deep Muscles of the Back

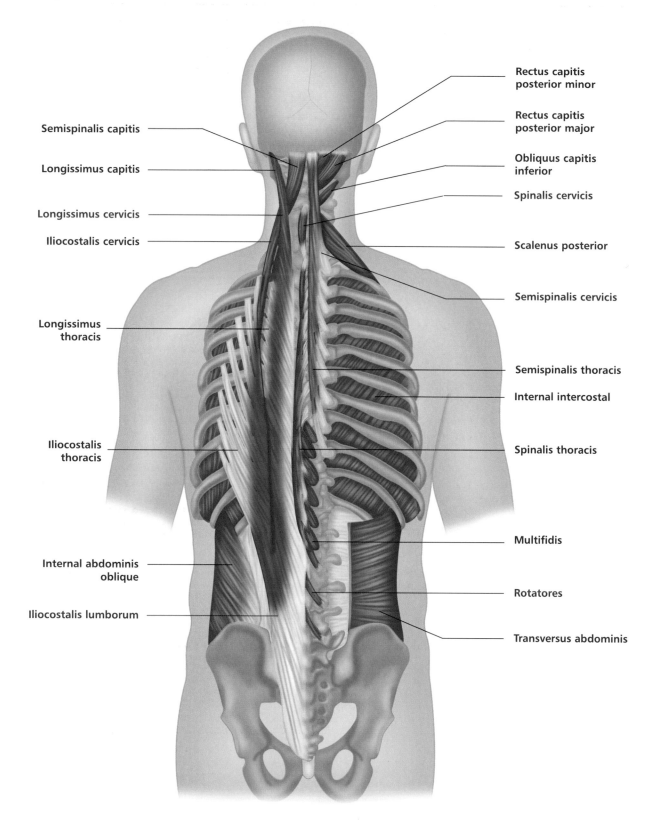

Semispinalis capitis

Longissimus capitis

Longissimus cervicis

Iliocostalis cervicis

Longissimus thoracis

Iliocostalis thoracis

Internal abdominis oblique

Iliocostalis lumborum

Rectus capitis posterior minor

Rectus capitis posterior major

Obliquus capitis inferior

Spinalis cervicis

Scalenus posterior

Semispinalis cervicis

Semispinalis thoracis

Internal intercostal

Spinalis thoracis

Multifidis

Rotatores

Transversus abdominis

Muscle	Origin	Insertion	Action	Nerve
Erector spinae (iliocostalis, longissimus, spinalis)	Slips of muscle arising from the sacrum. Iliac crest. Spinous and transverse processes of vertebrae. Ribs.	Ribs. Transverse and spinous processes of vertebrae. Occipital bone.	Extends and laterally flexes vertebral column. Helps maintain correct curvature of spine in the erect and sitting positions. Steadies the vertebral column on the pelvis during walking.	Dorsal rami of cervical, thoracic and lumbar spinal nerves.
Semispinalis (capitis, cervicis thoracis)	Transverse processes of cervical and thoracic vertebrae, (C1–T10).	Between nuchal lines of occipital bone and spinous processes of cervical and thoracic vertebrae, (C2–T4).	Capitis: Most powerful extensor of the head and assists in rotation. Cervicis and Thoracis: Extend thoracic and cervical parts of vertebral column. Assist rotation of thoracic and cervical vertebrae.	Dorsal rami of thoracic and cervical spinal nerves.
Rectus capitis posterior minor	Posterior tubercle of atlas.	Medial portion of inferior nuchal line of occipital bone.	Extends head.	Suboccipital nerve (dorsal ramus of first cervical nerve, C1).
Rectus capitis posterior major	Spinous process of axis.	Below lateral portion of inferior nuchal line of occipital bone.	Extends head. Rotates head to same side.	Suboccipital nerve (dorsal ramus of first cervical nerve, C1).
Multifidis	Posterior surface of sacrum, between the sacral foramina and posterior superior iliac spine. Mamillary processes of all lumbar vertebrae. Transverse processes of all thoracic vertebrae. Articular processes of lower four cervical vertebrae.	Parts insert into spinous process two to four vertebrae superior to origin; overall including spinous processes of vertebrae, (L5–C2).	Protects vertebral joints from movements produced by the more powerful superficial prime movers. Extension, lateral flexion and rotation of vertebral column.	Dorsal rami of spinal nerves.
Rotatores	Transverse process of each vertebra.	Base of spinous process of adjoining vertebra above.	Rotate and assist in extension of vertebral column.	Dorsal rami of spinal nerves.
Intercostales (externi, interni)	Externi: Lower eight ribs. Interni: Upper border of a rib and costal cartilage.	Externi: Upper border of rib below (fibres run obliquely forwards and downwards). Interni: Lower border of rib above (fibres run obliquely forwards and upwards towards the costal cartilage).	Muscles contract to stabilize the ribcage during various movements of the trunk. Prevents the intercostal space from bulging out or sucking in during respiration.	The corresponding intercostal nerves.
Obliquus internus abdominis	Iliac crest. Lateral two-thirds of inguinal ligament. Thoracolumbar fascia.	Bottom three or four ribs. Linea alba via an abdominal aponeurosis.	Compresses abdomen. Contraction of one side alone laterally bends and rotates the trunk.	Thoracic nerves, T7–T12, ilioinguinal and iliohypogastric nerves.
Transversus abdominis	Anterior two thirds of iliac crest. Lateral third of inguinal ligament. Thoracolumbar fascia. Costal cartilages of lower six ribs.	Linea alba via an abdominal aponeurosis, the lower fibres of which ultimately attach to the pubic crest and pecten pubis via the conjoint tendon.	Compresses abdomen, helping to support the abdominal viscera against the pull of gravity.	Ventral rami of thoracic nerves, T7–T12, ilioinguinal and iliohypogastric nerves.

Superficial Muscles of the Arm (Posterior View)

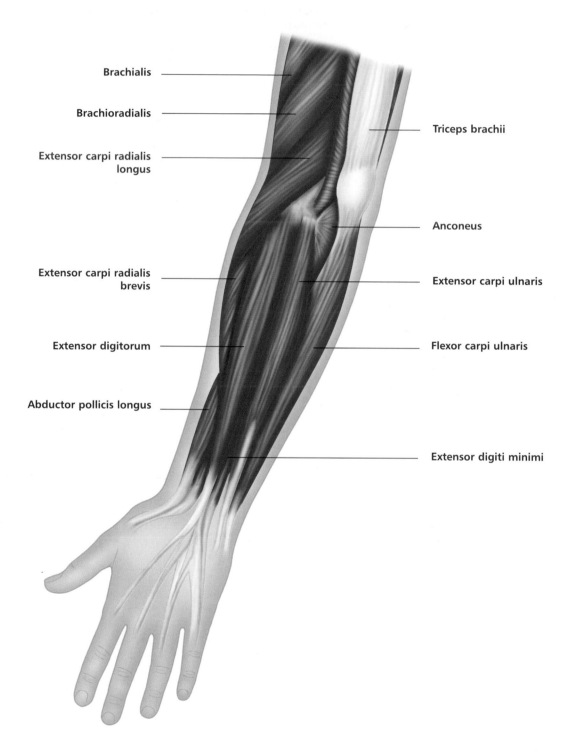

Brachialis

Brachioradialis

Extensor carpi radialis longus

Extensor carpi radialis brevis

Extensor digitorum

Abductor pollicis longus

Triceps brachii

Anconeus

Extensor carpi ulnaris

Flexor carpi ulnaris

Extensor digiti minimi

Muscle	Origin	Insertion	Action	Nerve
Extensor carpi radialis longus	Lower (distal) third of lateral supracondylar ridge of humerus.	Dorsal surface of base of second metacarpal bone, on its radial side.	Extends and abducts the wrist. Assists in flexion of the elbow.	Radial nerve, C5, **6**, **7**, 8.
Extensor carpi radialis brevis	Common extensor tendon from lateral epicondyle of humerus.	Dorsal surface of third metacarpal.	Extends wrist. Assists abduction of wrist.	Radial nerve, C5, **6**, **7**, 8.
Extensor digitorum	Common extensor tendon from lateral epicondyle of humerus.	Dorsal surfaces of all the phalanges of the four fingers.	Extends the fingers (metacarpophalangeal and interphalangeal joints). Assists abduction (divergence) of fingers away from the middle finger.	Deep radial (posterior interosseous) nerve, C6, **7**, **8**.
Abductor pollicis longus	Posterior surface of shaft of ulna, distal to the origin of supinator. Interosseous membrane. Posterior surface of middle third of shaft of radius.	Radial (lateral) side of base of first metacarpal.	Pulls metacarpal bone of the thumb into a position midway between extension and abduction (the tendon stands out during this movement). Abducts and assists in flexion of the wrist.	Deep radial (posterior interosseous) nerve, C6, **7**, **8**.
Triceps brachii	Long head: infraglenoid tubercle of the scapula. Lateral head: upper half of posterior surface of shaft of humerus (above and lateral to the radial groove). Medial head: lower half of posterior surface of shaft of humerus (below and medial to the radial groove).	Posterior part of the olecranon process of the ulna.	Extends elbow joint. Long head can adduct the humerus and extend it from the flexed position. Stabilizes shoulder joint.	Radial nerve, C6, **7**, **8**, T1.
Anconeus	Posterior part of lateral epicondyle of humerus.	Lateral surface of the olecranon process and upper portion of posterior surface of ulna.	Assists triceps brachii to extend forearm at elbow joint. May stabilize the ulna during pronation and supination.	Radial nerve, C7, **8**.
Extensor carpi ulnaris	Common extensor tendon from lateral epicondyle of humerus. Aponeurosis from mid-posterior border of ulna.	Medial side of base of fifth metacarpal.	Extends and adducts the wrist.	Deep radial (posterior interosseous) nerve, C6, **7**, **8**.
Extensor digiti minimi	Common extensor tendon from lateral epicondyle of humerus.	Extensor expansion of little finger with extensor digitorum tendon.	Extends little finger.	Deep radial (posterior interosseous) nerve, C6, **7**, **8**.

Superficial Muscles of the Arm (Anterior View)

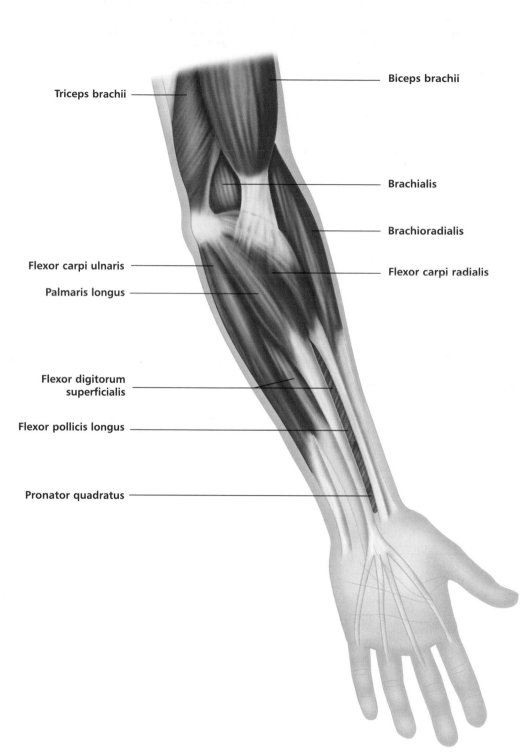

Triceps brachii

Biceps brachii

Brachialis

Brachioradialis

Flexor carpi ulnaris

Flexor carpi radialis

Palmaris longus

Flexor digitorum superficialis

Flexor pollicis longus

Pronator quadratus

Muscle	Origin	Insertion	Action	Nerve
Flexor carpi ulnaris	Humeral head: common flexor origin on the medial epicondyle of humerus. Ulnar head: medial border of olecranon. Posterior border of upper two-thirds of ulna.	Pisiform bone. Hook of hamate. Base of fifth metacarpal.	Flexes and adducts the wrist. May weakly assist in flexion of elbow.	Ulnar nerve, C7, **8**, T1.
Palmaris longus	Common flexor origin on the anterior aspect of the medial epicondyle of humerus.	Superficial surface of flexor retinaculum and apex of the palmar aponeurosis.	Flexes the wrist. Tenses the palmar fascia.	Median nerve, C(6), **7**, **8**, T1.
Flexor digitorum superficialis	Common flexor tendon on medial epicondyle of humerus. Coronoid process of ulna. Anterior border of radius.	Sides of the middle phalanges of the four fingers.	Flexes the middle phalanges of each finger. Can help flex the wrist.	Median nerve, C7, **8**, T1.
Biceps brachii	Short head: tip of corocoid process of scapula. Long head: supraglenoid tubercle of scapula.	Posterior part of radial tuberosity. Bicipital aponeurosis.	Flexes elbow joint. Supinates forearm. Weakly flexes arm at the shoulder joint.	Musculocutaneous nerve, C**5**, **6**.
Brachialis	Lower (distal) two thirds of anterior aspect of humerus.	Coronoid process of ulna and tuberosity of ulna.	Flexes elbow joint.	Musculocutaneous nerve, C**5**, **6**.
Brachioradialis	Upper two-thirds of the anterior aspect of lateral supracondylar ridge of humerus.	Lower lateral end of radius, just above the styloid process.	Flexes elbow joint. Assists in pronating and supinating forearm when these movements are resisted.	Radial nerve, C**5**, **6**.
Flexor carpi radialis	Common flexor origin on the anterior aspect of the medial epicondyle of humerus.	Front of the bases of the second and third metacarpal bones.	Flexes and abducts the carpus (wrist joint). Helps to flex the elbow and pronate the forearm.	Median nerve, C**6**, **7**, 8.

Deep Muscles of the Arm (Anterior View)

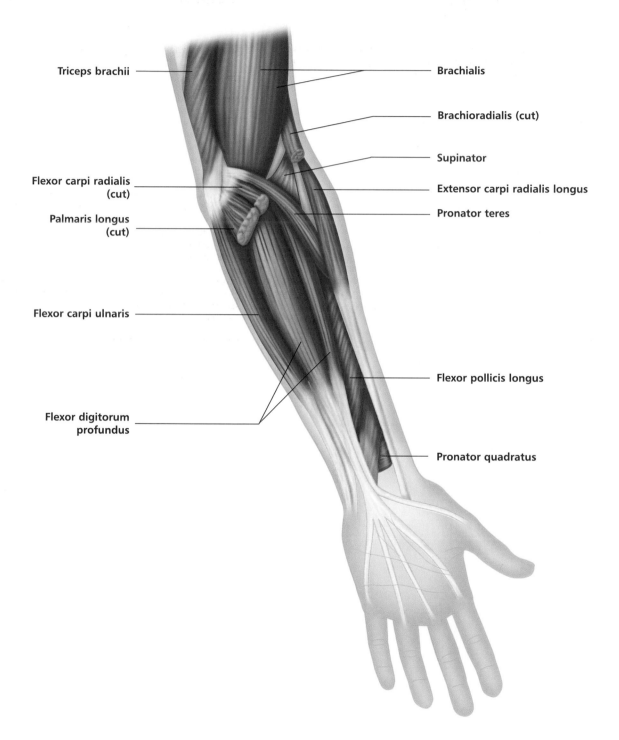

Triceps brachii

Flexor carpi radialis
(cut)

Palmaris longus
(cut)

Flexor carpi ulnaris

Flexor digitorum
profundus

Brachialis

Brachioradialis (cut)

Supinator

Extensor carpi radialis longus

Pronator teres

Flexor pollicis longus

Pronator quadratus

Muscle	Origin	Insertion	Action	Nerve
Flexor digitorum profundus	Upper two-thirds of the medial and anterior surfaces of the ulna, reaching up onto the medial side of the olecranon process. Interosseous membrane.	Anterior surface of base of distal phalanges.	Flexes distal phalanges (the only muscle able to do so). Helps flex all joints across which it passes.	Medial half of muscle: ulnar nerve, C7, **8**, T1. Lateral half of muscle: median nerve, C7, **8**, T1. Sometimes the ulnar nerve supplies the whole muscle.
Supinator	Lateral epicondyle of humerus. Radial collateral (lateral) ligament of elbow joint. Annular ligament of superior radio-ulnar joint. Supinator crest of ulna.	Dorsal and lateral surfaces of upper third of radius.	Supinates forearm.	Deep radial nerve, C5, **6**, (7).
Pronator teres	Humeral head: common flexor origin on the anterior aspect of the medial epicondyle of humerus. Ulnar head: coronoid process of the ulna.	Mid-lateral surface of radius.	Pronates forearm. Assists flexion of elbow joint.	Median nerve, C**6**, 7.
Flexor pollicis longus	Middle part of anterior surface of shaft of radius. Interosseous membrane. Medial border of coronoid process of ulna and/or medial epicondyle of humerus.	Palmar surface of base of distal phalanx of thumb.	Flexes the interphalangeal joint of the thumb. Assists in flexion of the metacarpophalangeal and carpometacarpal joints. Can assist in flexion of the wrist.	Anterior interosseous branch of median nerve, C(6), 7, **8**, T1.
Pronator quadratus	Distal quarter of anterior surface of shaft of ulna.	Lateral side of distal quarter of anterior surface of shaft of radius.	Pronates forearm and hand. Helps hold radius and ulna together, reducing stress on the inferior radio-ulnar joint.	Anterior interosseous branch of median nerve, C7, **8**, T1.

Superficial and Deep Muscles of the Hip and Thigh (Anterior View)

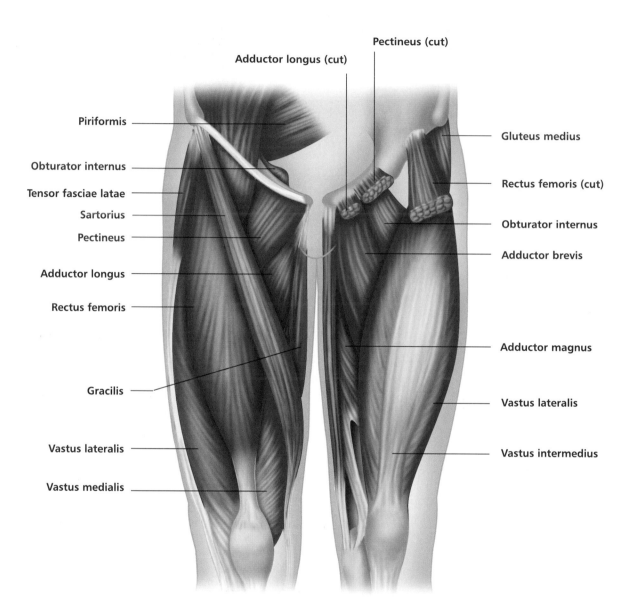

Pectineus (cut)

Adductor longus (cut)

Piriformis

Obturator internus

Tensor fasciae latae

Sartorius

Pectineus

Adductor longus

Rectus femoris

Gracilis

Vastus lateralis

Vastus medialis

Gluteus medius

Rectus femoris (cut)

Obturator internus

Adductor brevis

Adductor magnus

Vastus lateralis

Vastus intermedius

Muscle	Origin	Insertion	Action	Nerve
Piriformis	Internal surface of sacrum.	Superior border of greater trochanter of femur.	Laterally rotates hip joint. Abducts the thigh when hip is flexed. Helps hold head of femur in acetabulum.	Ventral rami of lumbar nerve, L(5) and sacral nerves, S**1**, **2**.
Sartorius	Anterior superior iliac spine and area immediately below it.	Upper part of medial surface of tibia, near anterior border.	Flexes hip joint. Laterally rotates and abducts the hip joint. Flexes knee joint. Assists in medial rotation of the tibia on the femur after flexion.	Two branches from the femoral nerve, L**2**, **3**, (4).
Pectineus	Pecten of pubis, between iliopubic (iliopectineal) eminence and pubic tubercle.	Pectineal line, from lesser trochanter to linea aspera of femur.	Adducts the hip joint. Flexes the hip joint.	Femoral nerve, L2, **3**, 4. Occasionally receives an additional branch from the obturator nerve, L3.
Rectus femoris	Anterior head: anterior inferior iliac spine. Posterior head: groove above acetabulum (on ilium).	Patella, then via patellar ligament to tuberosity of tibia.	Extends the knee joint and flexes the hip joint.	Femoral nerve, L**2**, **3**, **4**.
Vastus lateralis	Upper half of shaft of femur.	Lateral margin of patella, then via patellar ligament to tuberosity of tibia.	Extends the knee joint. Prevents flexion at knee joint.	Femoral nerve, L**2**, **3**, **4**.
Vastus medialis	Upper half of shaft of femur.	Medial margin of patella, then via patellar ligament to tuberosity of tibia. Medial condyle of tibia.	Extends knee joint. Prevents flexion at knee joint.	Femoral nerve, L**2**, **3**, **4**.
Vastus intermedius	Upper two-thirds of shaft of femur.	Deep surface of quadriceps tendon, then via patellar ligament to tuberosity of tibia.	Extends knee joint. Prevents flexion at knee joint.	Femoral nerve, L**2**, **3**, **4**.
Adductor longus	Anterior surface of pubis at junction of crest and symphysis.	Middle third of medial lip of linea aspera.	Adducts hip joint. Flexes extended femur at hip joint. Extends flexed femur at hip joint. Assists lateral rotation of hip joint.	Obturator nerve, L**2**, **3**, 4.
Adductor magnus	Inferior ramus of pubis. Ramus of ischium (anterior fibres). Ischial tuberosity (posterior fibres).	Whole length of femur.	Adducts and laterally rotates hip joint.	Obturator nerve, L2, **3**, **4**. Tibial portion of sciatic nerve, L**4**, 5, S1.
Adductor brevis	Outer surface of inferior ramus of pubis.	Lower two-thirds of pectineal line and upper half of linea aspera.	Adducts hip joint. Flexes extended femur at hip joint. Extends flexed femur at hip joint. Assists lateral rotation of hip joint.	Obturator nerve, (L2–L4).

Superficial and Deep Muscles of the Hip and Thigh (Posterior View)

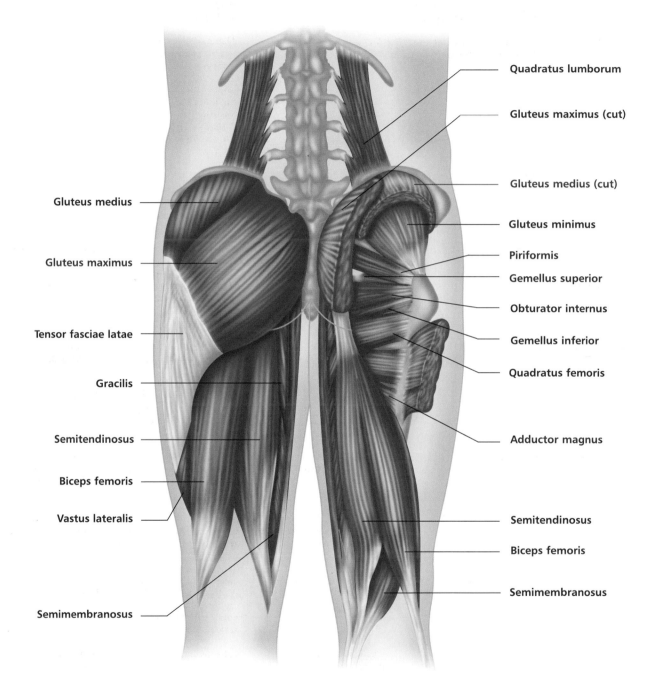

Gluteus medius

Gluteus maximus

Tensor fasciae latae

Gracilis

Semitendinosus

Biceps femoris

Vastus lateralis

Semimembranosus

Quadratus lumborum

Gluteus maximus (cut)

Gluteus medius (cut)

Gluteus minimus

Piriformis

Gemellus superior

Obturator internus

Gemellus inferior

Quadratus femoris

Adductor magnus

Semitendinosus

Biceps femoris

Semimembranosus

Muscle	Origin	Insertion	Action	Nerve
Gluteus maximus	Outer surface of ilium and posterior surface of sacrum and coccyx. Sacrotuberous ligament. Aponeurosis of erector spinae.	Gluteal tuberosity of femur. Iliotibial tract of fascia lata.	Extends and laterally rotates hip joint. Extends trunk. Assists in adduction of hip joint.	Inferior gluteal nerve, L5, S1, 2.
Gluteus minimus	Outer surface of ilium between anterior and inferior gluteal lines.	Anterior border of greater trochanter.	Abducts, medially rotates, and may assist in flexion of the hip joint.	Superior gluteal nerve, L4, 5, S1.
Gluteus medius	Outer surface of ilium.	Oblique ridge on lateral surface of greater trochanter of femur.	Abducts the hip joint. Anterior fibres medially rotate the hip joint. Posterior fibres slightly laterally rotate the hip joint.	Superior gluteal nerve, L4, 5, S1.
Tensor fasciae latae	Anterior part of outer lip of iliac crest, and outer surface of anterior superior iliac spine.	Joins iliotibial tract just below level of greater trochanter.	Flexes, abducts and medially rotates the hip joint. Tenses the fascia lata, thus stabilizing the knee.	Superior gluteal nerve, L4, 5, S1.
Gracilis	Lower half of symphysis pubis and inferior ramus of pubis.	Upper part of medial surface of shaft of tibia.	Adducts hip joint. Flexes knee joint. Medially rotates knee joint when flexed.	Anterior division of obturator nerve, L2, 3, 4.
Semitendinosus	Ischial tuberosity.	Upper medial surface of shaft of tibia.	Flexes and slightly medially rotates knee joint after flexion. Extends the hip joint.	Branches of sciatic nerve, L4, 5, S1, 2.
Biceps femoris	Ischial tuberosity. Back of femur.	Lateral side of head of fibula. Lateral condyle of tibia.	Flexes the knee joint. Extends the hip joint.	Branches of sciatic nerve, L5, S1, 2, 3.
Semimembranosus	Ischial tuberosity.	Posteromedial surface of medial condyle of tibia.	Flexes and slightly medially rotates knee joint after flexion. Extends the hip joint.	Branches of sciatic nerve, L4, 5, S1, 2.
Gemellus superior	External surface of ischial spine.	With tendon of obturator internus into medial surface of greater trochanter of femur.	Assists obturator internus to laterally rotate hip joint and abduct the thigh when the hip is flexed.	Nerve to obturator internus, lumbar nerve, L5, and sacral nerves, S1, 2.
Gemellus inferior	Upper margin of ischial tuberosity.	With tendon of obturator internus into medial surface of greater trochanter of femur.	Assists obturator internus to laterally rotate hip joint and abduct the thigh when the hip is flexed.	Branch of nerve to quadratus femoris, L4, 5, S1, (2).
Obturator internus	Inner surface of ischium, pubis and ilium.	Medial surface of greater trochanter of femur above trochanteric fossa.	Laterally rotates hip joint. Abducts the thigh when the hip is flexed. Helps hold head of femur in acetabulum.	Nerve to obturator internus, lumbar nerve, L5 and sacral nerves, S1, 2.
Quadratus femoris	Lateral border of ischial tuberosity.	Quadrate line that extends distally below intertrochanteric crest.	Laterally rotates hip joint.	A branch of lumbosacral plexus, L4, 5, S1, (2).
Quadratus lumborum	Posterior part of iliac crest. Iliolumbar ligament.	Twelfth rib. Transverse processes of lumbar vertebrae, (L1–L4).	Laterally flexes vertebral column. Helps extend lumbar part of vertebral column, and gives it lateral stability.	Ventral rami of the subcostal nerve and lumbar nerves, T12, L1, 2, 3.

Superficial Muscles of the Leg (Anterior View)

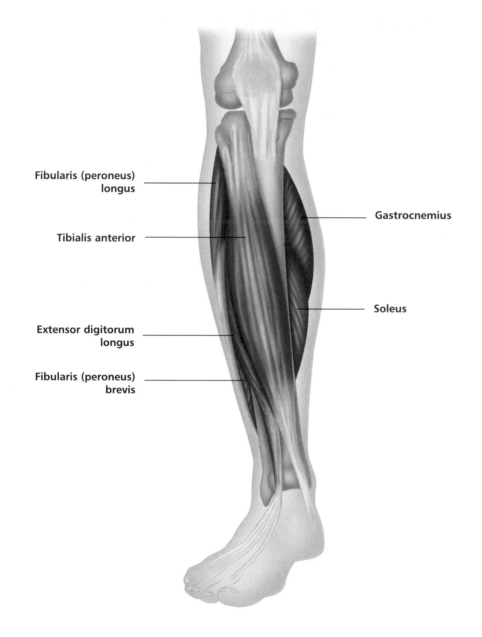

Fibularis (peroneus) longus

Tibialis anterior

Extensor digitorum longus

Fibularis (peroneus) brevis

Gastrocnemius

Soleus

Muscle	Origin	Insertion	Action	Nerve
Fibularis (peroneus) longus	Upper two-thirds of lateral surface of fibula. Lateral condyle of tibia.	Lateral side of medial cuneiform. Base of first metatarsal.	Everts the foot. Assists plantar flexion of ankle joint.	Superficial fibular (peroneal) nerve, L4, 5, S1.
Tibialis anterior	Lateral condyle of tibia. Upper half of lateral surface of tibia. Interosseous membrane.	Medial and plantar surface of medial cuneiform bone. Base of first metatarsal.	Dorsiflexes the ankle joint. Inverts the foot.	Deep peroneal nerve, L4, 5, S1.
Extensor digitorum longus	Lateral condyle of tibia. Upper two-thirds of anterior surface of fibula. Upper part of interosseous membrane.	Along dorsal surface of the four lateral toes. Each tendon dividing to attach to the bases of the middle and distal phalanges.	Extends toes at the metatarsophalangeal joints. Assists the extension of the interphalangeal joints. Assists in dorsiflexion of ankle joint and eversion of the foot.	Fibular (peroneal) nerve, L4, 5, S1.
Fibularis (peroneus) brevis	Lower two-thirds of lateral surface of fibula. Adjacent intermuscular septa.	Lateral side of base of fifth metatarsal.	Everts ankle joint. Assists plantar flexion of ankle joint.	Superficial fibular (peroneal) nerve, L4, 5, S1.
Gastrocnemius	Medial head: popliteal surface of femur above medial condyle. Lateral head: lateral condyle and posterior surface of femur.	Posterior surface of calcaneus (via the tendo calcaneus; a fusion of the tendons of gastrocnemius and soleus).	Plantar flexes foot at ankle joint. Assists in flexion of knee joint. It is a main propelling force in walking and running.	Tibial nerve, S1, 2.
Soleus	Upper posterior surfaces of tibia and fibula.	With tendon of gastrocnemius into posterior surface of calcaneus.	Plantar flexes ankle joint.	Tibial nerve, L5, S1, 2.

Deep Muscles of the Leg (Posterior View)

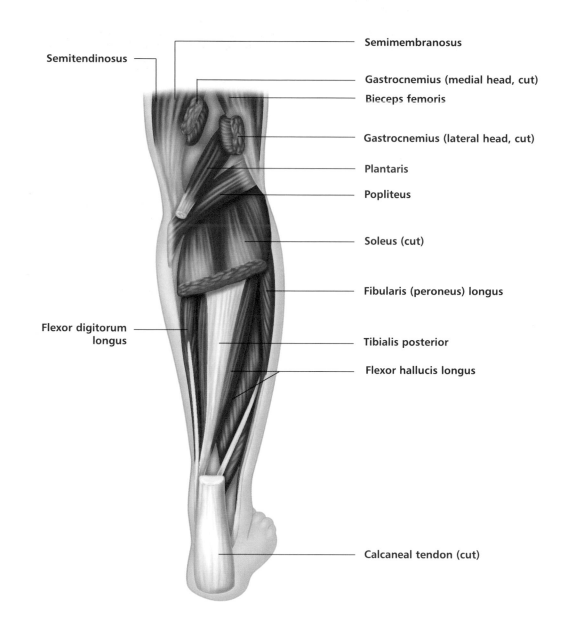

Semitendinosus

Semimembranosus

Gastrocnemius (medial head, cut)

Biceps femoris

Gastrocnemius (lateral head, cut)

Plantaris

Popliteus

Soleus (cut)

Fibularis (peroneus) longus

Flexor digitorum longus

Tibialis posterior

Flexor hallucis longus

Calcaneal tendon (cut)

Muscle	Origin	Insertion	Action	Nerve
Plantaris	Lower part of lateral supracondylar ridge of femur and adjacent part of its popliteal surface. Oblique popliteal ligament of knee joint.	Posterior surface of calcaneus.	Plantar flexes ankle joint. Feebly flexes knee joint.	Tibial nerve, L4, **5**, S**1**, (2).
Popliteus	Lateral surface of lateral condyle of femur. Oblique popliteal ligament of knee joint.	Upper part of posterior surface of tibia, superior to soleal line.	Laterally rotates femur on tibia when foot is fixed on the ground. Medially rotates tibia on femur when the leg is non-weight bearing. Assists flexion of knee joint. Helps reinforce posterior ligaments of knee joint.	Tibial nerve, L4, **5**, S1.
Tibialis posterior	Posterior surface of tibia and fibula. Most of interosseous membrane.	Tarsal bones and the second, third and fourth metatarsals.	Inverts the foot. Assists in plantar flexion of the ankle joint.	Tibial nerve, L(4), **5**, S1.
Flexor hallucis longus	Lower two-thirds of posterior surface of fibula. Interosseous membrane. Adjacent intermuscular septum.	Base of distal phalanx of great toe.	Flexes all the joints of the great toe. Helps to plantar flex the ankle joint and invert the foot.	Tibial nerve, L5, S**1**, **2**.
Flexor digitorum longus	Medial part of posterior surface of tibia, below soleal line.	Bases of distal phalanges of second through fifth toes.	Flexes all the joints of the lateral four toes. Helps to plantar flex the ankle joint and invert the foot.	Tibial nerve, L**5**, S**1**, (2).

An Introduction to the Anatomy Trains Myofascial Meridians

8

I am pleased to offer this addendum to Chris Jarmey's careful, concise, and beautifully illustrated survey of our structural anatomy. The following essay first outlines several metaphors helpful to a holistic approach to structural and movement therapies, and goes on to describe one map of larger functioning continuities within the musculo-skeletal system. These continuities, termed *myofascial meridians*, wind longitudinally through the soft tissues. These ideas are unfolded in greater detail in the book *Anatomy Trains* (Elsevier, 2001), and at www.AnatomyTrains.net.

Anatomy Trains provides a traceable basis for effective treatment at some distance from the site of dysfunction or pain. This new view of structural patterning also has far-reaching implications for treatment strategies, especially for long-standing postural imbalances, unsound body usage, and sequelae from injury or insult.

Practical Holism

The very structure of our language, and its cause-and-effect epistemology, requires that we understand any system by dividing it into its constituent parts, in order to define the contribution of each identifiable bit to the whole. A tree has roots, trunk, branches and leaves, each with an essential function.

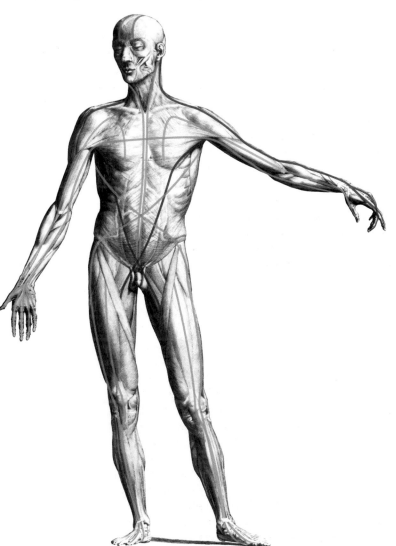

The leaves have stomata, mesophyll, and veins. The veins have xylem and phloem bundled in a sheath. And so on down the line of smaller and smaller building blocks: cells, macromolecules, atoms, and quantum forces. This analytical process is fundamental to our Western comprehension of the world. But this way of thinking presents one significant danger when we apply it to living systems such as trees and ourselves: the tree did not glue a root system to a trunk and bolt on branches with leaves wired to them. It sprang from a single seed, and is ever and always a co-evolving unitary set of system interactions from root to leaf. In reality, the parts are never separated, and are always codependent.

Humans are not assembled out of parts like a car or a computer. 'Body as machine' is a useful metaphor, but like any poetic trope, it does not tell the whole story. In our modern perception of human movement anatomy, however, we are in danger of making this metaphor into the be all and end all. In actual fact, our bodies are conceived as a whole, and grow, live, and die as a whole – but our mind is a knife (*see* figure 8.1).

Figure 8.1: The Anatomy Trains map of myofascial connections.

The medical idea that we are assembled from pieces – an easily-swallowed idea in this age of surgical 'spare parts', both mechanical and human – stems from Aristotle's premises, but really took hold in the study of humans when the body was first anatomized by Andreas Vesalius, Albinus, and other courageous explorers of the late Renaissance. The tool of choice, as implied in the very word anatomy, was a blade. From the flint cleaver to the laser scalpel, the animal and human body has been divided along finer and finer lines. Later, Cartesian dualism described the body as a 'soft machine', and students of anatomy and physiology used reductionistic mechanism to go about explaining the role of each identifiable part. The various physical laws of Isaac Newton further cemented our place within the mechanical universe. What were glorious and liberating ideas in their own time, however, have become imprisoning, restrictive concepts in ours (*see* figure 8.2).

How do our 'parts' arise? The human body stems from a single fertilized human ovum, which proliferates wildly. The daughter cells then specialize as outlined in Chapter 2. Each tissue cell exaggerates some function of the ovum and cells in general – e.g. a muscle specializes in contraction, a neuron in conduction, epithelia in secretion, etc. – and conversely other functions diminish. A nerve cell conducts extraordinarily well, but as a result of that specialization cannot easily reproduce itself. Epithelia do very well at creating enzymes, but lose the ability to significantly contract. Yet each cell still partakes of the unique individual whole in its constant communication with its neighbors, near and far, and in the similarities of chemical structure, from glucose as a universal fuel right on down to the tangled helix of DNA (*see* figure 8.3).

Before specific cuts are made, what we are pleased to call 'the brain' never exists as an entity separate from its connective tissue surroundings, its blood supply, and the peripheral and autonomic nerves that extend the brain throughout the body. The 'biceps brachii' can only exist as a separate structure with a knife's intervention to divide its ends from various attachments, its connections with surrounding myofascial units such as the brachialis, as well as its nerve and blood supply, without which it simply could not function. The idea that there are separate parts – a liver, a brain, a biceps – may be the way that we think, but it is not the way physiology 'thinks'.

Figure 8.2: Vesalius's woodcuts from 1548 show both the origami layering and the directional 'grain' of the myofascial system.

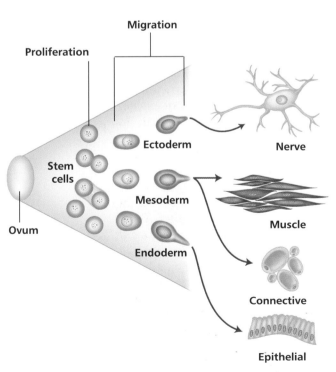

Figure 8.3: From the generalized ovum, cells proliferate, migrate, and differentiate into functionally specialized tissues.

The Single Muscle Theory

This image of 'separate' muscles – muscles as parts – leads to the prevalent method of analyzing muscle action, which is employed frequently (and to good purpose) throughout this and most other atlases: "Imagine that the skeleton were denuded of all but the given muscle; what would that single muscle do to the skeleton acting on its own?" Call this the 'single muscle' theory.

In this single muscle theory, the biceps gets defined as a radio-ulnar supinator, an elbow flexor, and a weak flexor of the shoulder (*see* figure 8.4a). In the Anatomy Trains view, additional information is added to this: "The biceps brachii is an element in a continuous fascial plane or myofascial meridian which runs from the outside of the thumb to the 4th rib and beyond." The second statement does not negate the first, but it adds a context for understanding the biceps' role in stabilizing the thumb (down the myofascial line), and keeping the chest open and the breath full (up the line) (*see* figure 8.4b).

This 'body as assembled machine' idea is so pervasive – and as in this book, the maps based on this perspective are so understandable and useful – that it is difficult to think outside its parameters. Thinking in 'wholes', attractive as it is to contemporary holistic therapists, simply has yet to lead to useful maps. The 'everything is connected to everything else' philosophy expounded in our opening paragraphs, while actually technically accurate, leaves the practitioner adrift in this sea of connections, unsure as to whether that frozen shoulder will respond to work in the elbow, the contralateral hip, or to a reflex point on the ipsilateral foot. While any of these might work, useful maps are necessary to organize our therapeutic choices into something better than a guess.

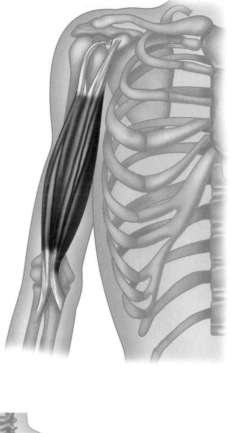

a)

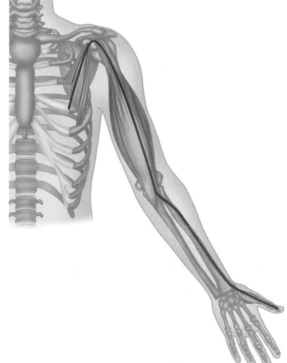

b)

Figure 8.4: a) the biceps brachii considered as a separate muscle, b) the biceps brachii is also part of a longitudinal myofascial continuity.

In short, we know the body is interconnected on many levels, but we need better treatment strategies than 'press and pray'. What can we learn when we shift from a 'symptom-oriented' view of the body to a 'system-oriented' one?

This myofascial meridians concept provides such a map of the structural body, a map that provides a practical transition between the individual 'parts' that the authors have so brilliantly catalogued herein, and the 'whole' of a human being, a gestalt of physics, physiology, stored experience, and current awareness which defies mapping. This intermediate map of the body's locomotor fabric opens up new avenues of treatment consideration, particularly for stubborn chronic conditions and global postural effects.

Whole Body Communicating Networks

Central to this new 'Anatomy Trains' map is the functional unity of the connective tissue system. There are exactly three networks within the body which, magically extracted intact, would show us the shape of the whole body, inside and out: the neural net, the vascular system, and the extracellular fibrous web created by the connective tissue cells (*see* figure 8.5).

Large communities of cells need vast infrastructure to survive in crowded conditions, especially out of the water, resting on land, and surrounded by air. It is an astounding feat of engineering we take for granted every day: a community of 70 trillion diverse, humming and semi-autonomous cells, each built for undersea living, organizes itself to get up and walk around, while simultaneously providing each cell with a mechanically stable environment, oceanic conditions of chemical exchange, and the information it needs to participate meaningfully in the day's work.

Every living cell needs to be within four cell layers or so of the fluid exchange provided by the vascular system. Without the ability to deliver chemistry to, and suck waste from, every single body region, the underserved area becomes stressed, then distressed, and will finally shrivel or burst and die, as happens in necrotic or gangrenous tissue. Secondly, every cell needs to be within reach of the nervous system to regulate its

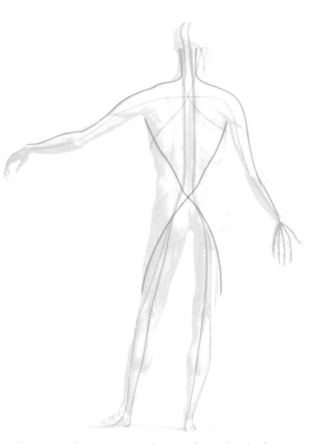

Figure 8.5: The Anatomy Trains map (posterior view).

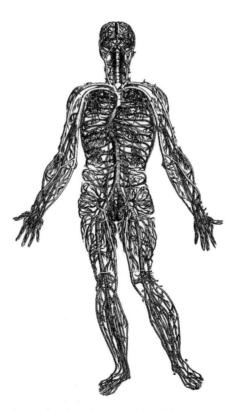

Figure 8.6: Vesalius's version of the vascular net.

activity with other cells in other areas of the body. And every cell needs to be structurally held in place (or directed in a flow, in the case of blood and other mobile cells) by the connective tissue net. Any given single cubic centimeter – let alone Shylock's pound of flesh – taken from the body would contain elements of all three of these nets – neural, vascular, and fascial. Seen in such a systemic way, the idea of the body as simply a collection of parts begins to lose its luster. We all survive because we are an interwoven set of systems (*see* figures 8.6 & 8.7).

Each of these nets is networked across the entire body from central organizing structures. From its central spinal cord with the bulbous brain plexus at one end, the nervous system spreads throughout the body in the familiar radiating pattern seen here, all to form a simulation of our inner and outer world and a coordinated behavioral response. The circulatory system, from the axis of the aorta and the muscular heart, links the thousands of miles of capillaries and lymphatics in a circle that serves the chemical needs of the cellular community. And from the central armature of the notocord (the embryological form of the vertebral bodies and discs), connective tissue spreads out to create protective sacs and nets around all the cells, structures, and systems of the body, organizing stable mechanical relationships, allowing certain movements and discouraging others.

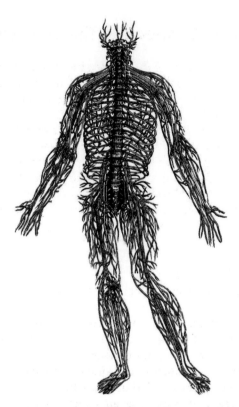

Figure 8.7: Vesalius's version of the neural net.

All three networks communicate throughout the body. The nerves carry sensory data in, construct a second-to-second picture of the world, and send signals out to the muscles and glands, at speeds between 7 and 170 miles per hour. The fluid systems circulate chemistry around the body every few minutes, though many chemical rhythms fluctuate in hourly, daily, or as women know, monthly cycles. The fibrous system communicates mechanical information – tension and compression – via the intercellular matrix of fascia, tendon, ligament, and bone. This information is a vibration that travels at the speed of sound, about 700 mph – slower than light, but definitely faster than the nervous system. The speed of plastic deformation and compensation in the connective tissue system, however, is measured in weeks, months, and years. Thus the fibrous system is both the fastest (in communicating) and the slowest (in responding) of the three.

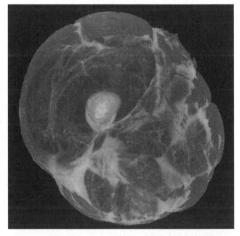

Figure 8.8: A section of the thigh, viewed from above, with all other tissues but the connective tissues removed. Courtesy of Jeff Linn, derived from the Visible Human Data Project.

Unlike the neural and vascular systems, this connective tissue net has yet to be well mapped, because it is considered to be the 'dead' material that we need to remove to see the 'interesting' neural, vascular, muscular, and other local systems. Because the connective tissue provides the divisions along which the scalpel runs to parse out other systems, the connective tissue has also been studied less as a system than other more familiar systems.

So, as a thought experiment: what if, instead of dividing the body into individual identifiable structures, we were to dip it into a solvent that stripped away all the cellular material but left the entire extracellular matrix intact (*see* figure 8.8)?

The Connective Tissue System

This system of the connective tissue matrix can be seen as our 'organ of form'. From the moment of the first division of the ovum, the intercellular matrix of the connective tissue exists as a secreted glycoaminoglycans (mucous) gel that acts to glue the cells together. Around the end of the second week of embryological development, the first fibrous version of this net appears, a web of fine reticulin spun by specialized mesodermal cells on either side of the developing notocord (spine). This net is the origin of our fascial web – our 'metamembrane' – the singular container that shapes our form and directs the flow of all our biochemical processes (*see* figure 8.9). The ability of the connective tissue cells to alter and mix the three elements of the intercellular space – the water, the fibers, and the gluey ground substance gel of glycoaminoglycans – produces on demand the wide spectrum of familiar building materials in the body – bone, cartilage, ligament, tendon, areolar and adipose networks – all the varieties of biological fabric. The body's joints, the 'organ system of movement', are almost entirely composed of extracellular matrix constructed by the various connective tissue cells.

The predominant scientific view is that this extracellular matrix material is *'non-living'*, but is this in fact an accurate description? The hydrophobic fibrous network of collagen, elastin, and reticulin, and the hydrophilic gelatinous ground substance is clearly extracellular, in that it is all manufactured inside a connective tissue cell and then extruded out into the intercellular space, where it may lose all contact with its original producing cell. The fiber-gel matrix remains an immediate part of the environment of every cell, like cellulose in plants, or the coral's limestone apartment building.

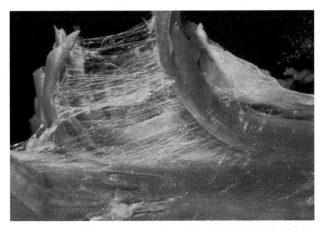

Figure 8.9: A magnification of myofascial tissue – individual muscle fibers within the cotton floss of the endomysial fascia. Photograph courtesy of Ron Thompson.

Given, however, that the animal (and human) extracellular matrix is so responsive to change – some in passive response to outside forces, some in active cellular response to damage or need – and given that it is a liquid crystal capable of storing and transmitting information, and finally given that it is so intimately married to the lives of our cells, I choose to think of it as living. It is part of our adaptive response to the needs of practical continuance; it is part of the very fabric of our consciousness. Of course the point is debatable, but for the rest of this essay we take a mildly vitalistic stance that includes this extracellular matrix as belonging in and partaking of the field of the living being.

This extracellular matrix, taken as a whole, not only unites the various elements of the body; in large part it unites the many branches of medicine. We leave the description of its physiology and biochemistry to others more familiar with it. Here we concentrate on two of aspects of the matrix's spatial configuration: its 'double bagging' arrangement and seeing the interplay of its elements in terms of tensegrity geometry.

The Double-Bag Theory

If we move through embryology from the initial development of the connective tissue metamembrane to watch its further elaboration, what we see is a fabulous demonstration of origami. The involution of gastrulation and the subsequent lateral and sagittal folds are followed by literally thousands of others, taking the original simple 3-dimensional spider-web of surrounding and investing fascia and folding it into more than a thousand compartments and divisions that we subsequently cut out and identify as separate 'parts'. Initial folds create the dorsal cavity for the brain, and ventral cavity for the organs, and surround each organ with a double-layered fascial sac. One of the final folds brings the two halves of the palate together, which explains why a cleft palate is such a common birth defect.

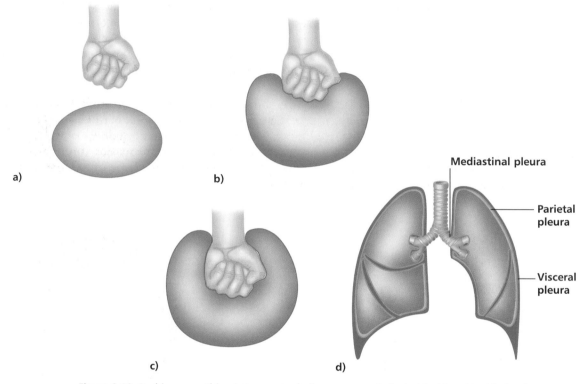

Figure 8.10: Pushing something into a water balloon surrounds that object in a 'double-bag', with lubrication in between the two layers.

Pushing a fist into a water balloon produces a doubled bag around the hand. It is actually one sac, but pressure gradients push into the sac, involuting on itself like a sock turned halfway inside-out. When organ tissue cells are similarly 'pushed' by the forces of embryological growth into a connective tissue bag like the coelom, this likewise creates the appearance of two sacs separated by lubricating fluid: an inner one closely adherent to the organ (usually softer), and another (usually tougher) around the outside (*see* figure 8.10). This method of biological double-bagging surrounds the heart with a soft connective tissue endocardium and a tough outer pericardium. Similarly, the lungs have the visceral pleura and parietal pleura, and the intestines lie in the mesentery and the peritoneum, and the brain has the pia and dura mater. Our thesis here: the musculo-skeletal system is similarly organized within a fascial 'double-bag'.

Imagine a water balloon half-filled with an electrically-sensitive jelly (muscle). Put a couple of cylindrical objects on it, like short dowels or marker pens (bones), end to end. Now push the cylinders into the balloon until it comes up around them and encloses them (*see* figure 8.11).

Your musculo-skeletal system, seen as a whole, is wrapped up something like this. The part adhering to the cylinders is like the periosteum around the bones. In the space between the 'bones', this part of the balloon is like the ligamentous capsule of the joints, which thickens into specific ligaments as required by the forces acting on it. The outer part of the balloon is like the superficial fascial leotard (or, in medical parlance, the deep investing fascia). The part of the balloon where the two ends of the balloon meet (as in the right side of figure 8.11b) represents the intermuscular septa, which are similarly double layered walls that run from the superficial fascia to the periosteum.

Seen in this way, the periosteum-joint capsule 'inner bag' is filled with hard bone alternating very soft joint tissues and fluid. The outer bag is filled with muscle and varying densities of fascia. In order to make the two bags interact successfully to move us around, we simply come along with a soldering iron and heat seal the outer bag down onto the inner in specific places, making what we call a muscle attachment. With this image, we get away from the shibboleth that there are some six hundred muscles in the body. There is, in fact, only one muscle. One mind, and one muscle – it just hangs around in six hundred pockets within the unitary fascial bag. It is a struggle, I know, to rise above this concept of individual muscles, but the view from the mountain of wholeness is breathtaking.

Seen through this lens, the myofascial meridians simply track the warp and weft through this outer bag of myofascial webbing. Where does this webbing continue in straight lines, lines that can transmit forces which travel out from their local areas to create global effects via the interconnectedness of this overall double bag? The answer to this question provides a map for tracking strain transmission throughout this system. How we move and are moved during our lives shapes this web; the shape of the web in turn helps to determine our experience of living in our bodies.

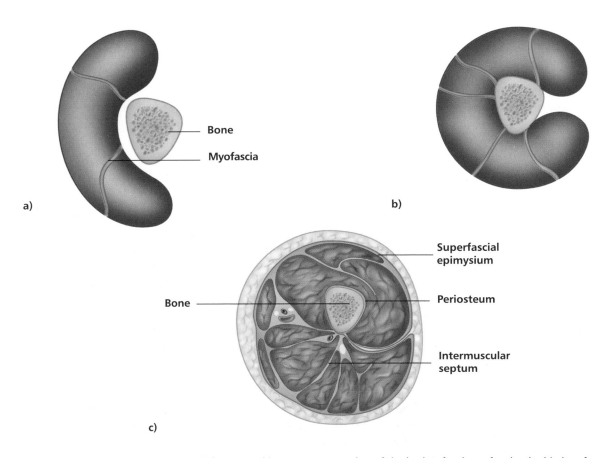

Figure 8.11: The relation between myofasciae and bone seen as another of the body's fondness for the double-bag form.

Tensegrity

One other holistic image is necessary to jump out of this 'machine made out of parts' image so ingrained in our systems – tensegrity geometry. The normal geometric picture of our anatomy is that the skeleton is a continuous compression framework, like a crane or a stack of blocks, and the muscles hang from it like the cables. This leads to the single muscle theory again – the skeleton is stable but moveable, and we parse out what each muscle does to that framework on its own, adding them together to analyze the movement. A little thought, however, soon puts this idea out to pasture. Take the muscles away, and the skeleton is anything but stable; take all the soft-tissue away and the bones would clatter to the floor, as they do not interlock or stack in any kind of stable way.

If we can get away from the idea that bones are like girders and muscles are the cables that move the girders, we are led to a class of structures called 'tensegrity' (the integrity lies in the balance of tension) (*see* figure 8.12). Originated by Kenneth Snelson and developed by Buckminster Fuller, tensegrity geometry more closely approximates the body as we live and feel it than does the old 'crane' model. In the dance of stability and mobility that is a human moving, the bones and cartilage are clearly compression-resisting struts that push outward against the myofascial net. The net, in turn, is always tensional, always trying to pull inward toward the center. Both elements are necessary for stability, and both contribute to practical mobility.

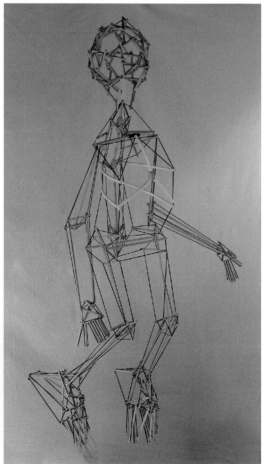

In this new orthopedic model, the boney struts 'float' within the sea of tension provided by the soft tissues. The position of the bones is thus dependent on the tensional balance among these soft-tissue elements. This model is of great importance in seeing the larger potential of soft-tissue approaches to structure, in that boney position and posture is far more dependent on soft-tissue balance than on any high-velocity thrusting of bones back into 'alignment'.

In this view, much expanded in our other writings, the Anatomy Trains Myofascial Meridians map the global lines of tension that traverse the entire body's muscular surface, acting to keep the skeleton in shape, guide the available tracks for movement, and coordinate global postural patterns. Research supports the idea of tensegrity geometry ruling mechanical transmission from the cellular level on up, and macro-level models, such as the one pictured here, are becoming more anatomically accurate.

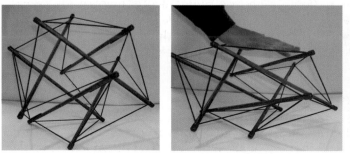

Figure 8.12: Tensegrity structures, when stressed, tend to distribute rather than concentrate strain. The body does the same, with the result that local injuries soon become global strain patterns.

For more on tensegrity: www.kennethsnelson.net, http://web1.tch.harvard.edu/research/ingber/Tensegrity.html,
www.intensiondesigns.com, www.biotensegrity.com

The Anatomy Trains: Rules of the Road

Let us now step down to intermediate level, somewhere between these overarching global considerations and the useful detailed anatomy that comprises the rest of this book. The concept is very simple: if we follow the grain of the fascial fabric, we can see where muscles are linked up longitudinally. When this is done, there are 12 or so major myofascial meridians that appear, forming clear lines, or tracks, that traverse the body. A few rules and terminological considerations apply to their construction.

1. Myofascial continuities must run in straight lines.

Since an Anatomy Train is a line of tensional pull, not compressional push, it must therefore travel in a more or less straight line. So the first and major rule is that adjacent myofascial structures must 'line up' in the fascia, without major changes of direction or depth. While the hamstrings and gluteus maximus might be functionally connected in running or climbing stairs, the change in direction and change of level between the two prevent them from being a fascial continuity. The hamstrings and erector spinae, however, are clearly connected in a straight-ish line via the sacrotuberous ligament.

2. Fascia is continuous while muscles are discrete – 'tracks' and 'stations'.

The stripes of muscles and fascia are termed 'tracks', and what are commonly known as muscles attachments are termed 'stations', to emphasize the continuous nature of the fascial fabric. Muscles themselves may be discrete, but the fascia that contains them is continuous, and communicates to the next structure up or down the line. The external oblique muscle and the serratus anterior muscle may be separate and have separate functions, but the sinew that envelops each of them is part of the same fascial plane, which communicates across their attachment points and beyond.

3. Fascial planes can divide or blend – 'switches' and 'roundhouses'.

Fascial planes sometime split into two planes, or conversely, two planes blend into one. We call these dividing places 'switches' (UK: points); and the physics of the situation will determine which plane takes how much of the force involved. The rhomboids thus communicate fascially with both the serratus anterior and/or the rotator cuff, depending on the position, load, and orientation of these and surrounding structures.

Places where many muscles meet and provide competitive directions for where a bony structure might be pulled – the ASIS for instance, or the scapula – are termed 'roundhouses'.

4. Deeper, single-joint muscles hold posture – 'expresses' and 'locals'.

Monoarticular, or one-joint, muscles are termed 'locals', whereas multiarticular muscles are termed 'expresses'. Our finding is that posture is more often held in the deeper single-joint locals, and not in the coordinating expresses which often overlie them. Thus, one looks to relieve a chronically flexed hip via the iliacus or pectineus myofascia, more often than the rectus femoris or sartorius, both of which also cross the knee.

5. When the rules get bent – 'derailments'.

And finally, we sometimes encounter 'derailments', where the myofascial meridian does not utterly conform to the above rules, but works under particular conditions or positions. For instance, the line of fascia along the back of the body is a continuous string of fascia when the knee is straight, but 'de-links' into two pieces – one above and one below, when the knee is significantly flexed. This explains why nearly every classic yoga stretch for the Superficial Back Line, as we term it, has the knees in the extended position, and why it is easier to pick up your dropped keys with even slightly flexed knees than with straight legs.

Individual myofascial meridians can be viewed as one-dimensional tensional lines that pass from attachment point to attachment point from one end to the other. They can be viewed as two-dimensional fascial planes that encompass larger areas of superficial fasciae. Or they can be seen, as they are here as three-dimensional set of muscles and connective tissues, which, taken together, comprise the entire volume of the musculo-skeletal system.

Summary of the Lines

With these rules in mind, we can construct 12 myofascial meridians in common use in human stance and movement:

- Superficial Front Line
- Superficial Back Line
- Lateral Line (2 sides)
- Spiral Line
- Arm Lines (4)
- Functional Lines (2 – front and back)
- Deep Front Line

The first three lines are termed the 'cardinal' lines, in that they run more or less straight up and down the body in the four cardinal directions – front, back, and left and right sides.

Superficial Front Line

The Superficial Front Line (SFL) runs on both the right and left sides of the body from the top of the foot to the skull, including the muscles and associated fascia of the anterior compartment of the shin, the quadriceps, the rectus abdominis, sternal fascia, and sternocleidomastoideus muscle up onto the galea aponeurotica of the skull. In terms of muscles and tensional forces, the SFL runs in two pieces – toes to pelvis, and pelvis to head, which function as one piece when the hip is extended, as in standing (*see* figure 8.13).

In the SFL, fast-twitch muscle fibers predominate. The SFL functions in movement to flex the trunk and hips, to extend the knee, and to dorsiflex the foot. In standing posture, the SFL flexes the lower neck but hyperextends the upper neck. Posturally, the SFL also maintains knee and ankle extension, protects the soft organs of the ventral cavity, and provides tensile support to lift those parts of the skeleton which extend forward of the gravity line – the pubes, the ribcage, and the face. And, of course, it provides a balance to the pull of the Superficial Back Line.

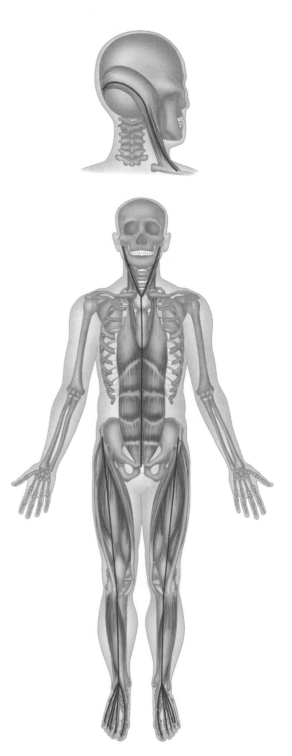

Figure 8.13: The Superficial Front Line (SFL).

A common human response to shock or attack, the startle response, can be seen as a shortening of the SFL. Chronic contraction of this line – common after trauma, for example – creates many postural pain patterns, pulling the front down and straining the back.

Superficial Back Line

The Superficial Back Line (SBL) runs from the bottom of the toes around the heel and up the back of the body, crossing over the head to its terminus at the frontal ridge at the eyebrows. Like the SFL, it also has two pieces, toes to knees and knees to head, which function as one when the knee is extended. It includes the plantar tissues, the triceps surae, the hamstrings and sacrotuberous ligament, the erector spinae, and the epicranial fascia.

The SBL functions in movement to extend the spine and hips, but to flex the knee and ankle. The SBL lifts the baby's eyes from primary embryological flexion, progressively lifting the body to standing (*see* figure 8.14).

Posturally, the SBL maintains the body in standing, spanning the series of primary and secondary curves of the skeleton (including the cranium and heel in the catalogue of primary curves, and knee and foot arches in the list of secondary curves). This results in a more densely fascial line than the SFL, with strong bands in the legs and spine, and a predominance of slow-twitch fibers in the muscular portion.

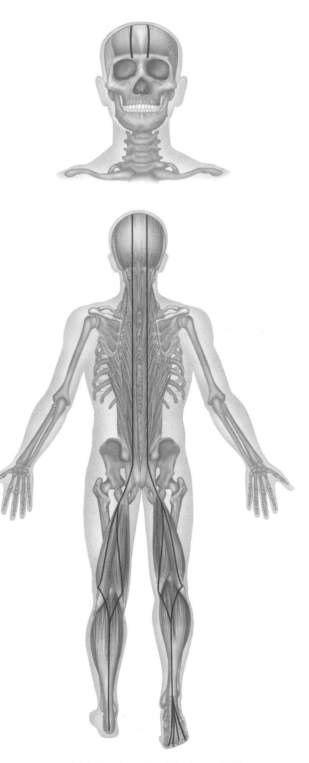

Figure 8.14: The Superficial Back Line (SBL).

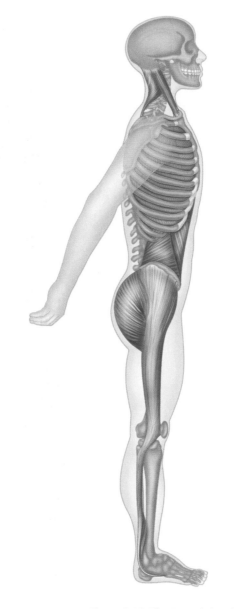

Figure 8.15: The Lateral Line (LL).

Lateral Line

The Lateral Line (LL) traverses each side of the body from the medial and lateral midpoints of the foot around the fibular malleolus and up the lateral aspects of the leg and thigh, passing along the trunk in a woven pattern that extends to the skull's mastoid process (*see* figure 8.15).

In movement, the LL creates lateral flexion in the spine, abduction at the hip, and eversion at the foot, and also operates as an adjustable 'brake' for lateral and rotational movements of the trunk.

The LL acts posturally like tent guy-wires to balance the left and right sides of the body. Also, the LL contains more than creates movement in the human, directing the flexion-extension that characterizes our direction through the world, restricting side-to-side movement that would otherwise be energetically wasteful.

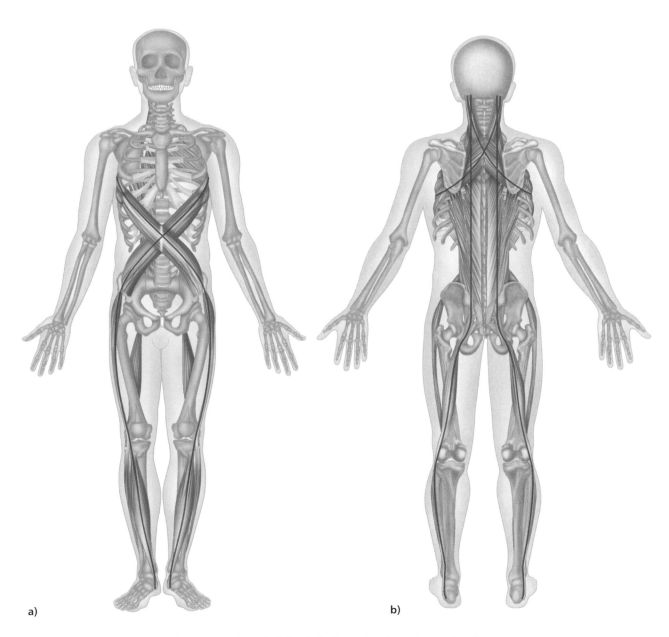

a)   b)

Figure 8.16: The Spiral Line (SL); a) anterior view, b) posterior view.

Spiral Line

The Spiral Line (SL) winds through the three cardinal lines, looping around the trunk in a helix, with another loop in the legs from hip to arch and back again. It joins one side of the skull across the midline of the back to the opposite shoulder, and then across the front of the torso to the same side hip, knee and foot arch returning up the back of the body to the head (*see* figure 8.16).

In movement, the SL creates and mediates rotations in the body. The SL interacts with the other cardinal lines in a multiplicity of functions.

In posture, the SL wraps the torso in a double helix that helps to maintain spinal length and balance in all planes. The SL connects the foot arches with tracking of the knee and pelvic position. The SL often compensates for deeper rotations in the spine or pelvic core.

Arm Lines

- Superficial Front Arm Line
- Deep Front Arm Line
- Superficial Back Arm Line
- Deep Back Arm Line

The four Arm Lines run from the front and back of the axial torso to the tips of the fingers. They are named for their planar relation in the composition of the shoulder, and roughly parallel the four lines in the leg. These lines connect seamlessly into the other lines particularly the Lateral, Functional, Spiral, and Superficial Front Lines (*see* figure 8.17).

In movement the arm lines place the hand in appropriate positions for the task before us – examining, manipulating, or responding to the environment. The Arm Lines act across 10 or more levels of joints in the arm to bring things to us or to push them away, to push, pull, or stabilize our own bodies, or simply to hold some part of the world still for our perusal or modification.

The Arm Lines affect posture indirectly, since they are not part of the structural column. Given the weight of the shoulders and arms, however, displacement of the shoulders in stillness or in movement will affect other lines. Conversely, structural displacement of the trunk in turn affects the arms' effectiveness in specific tasks and may predispose them to injury.

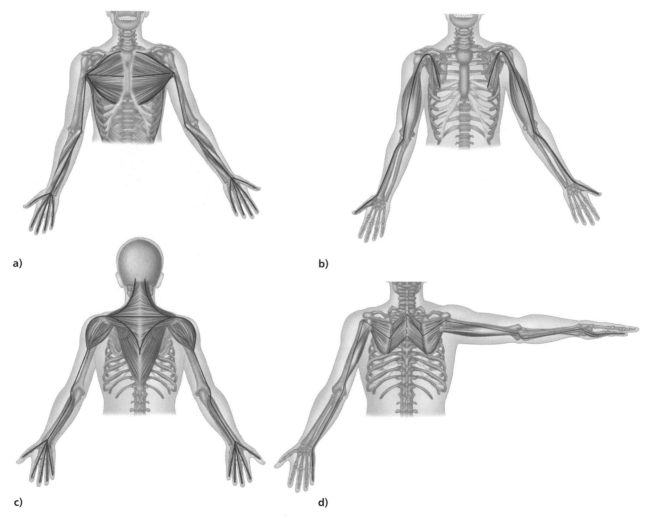

a)

b)

c)

d)

Figure 8.17: The four Arm Lines;
a) Superficial Front Arm Line, b) Deep Front Arm Line, c) Superficial Back Arm Line, d) Deep Back Arm Line.

Beyond the straightforward progression of the meridians from the trunk to the four corners of the hands, there are many 'crossover' muscles that link these lines to ether, providing additional support and stability for the extra mobility the arms have relative to the legs.

Functional Lines

• Front Functional Line
• Back Functional Line

The two Functional Lines join the contralateral girdles across the front and back of the body, running from one humerus to the opposite femur and vice versa (*see* figure 8.18).

The Functional Lines are used in innumerable movements, from walking to the most extreme sports. They act to extend the levers of the arms to the opposite leg as in a kayak paddle, a baseball throw or a cricket pitch (or vice versa in the case of a football kick). Like the Spiral Line, the Functional Lines are helical, and thus help create strong rotational movement. Their postural function is minimal.

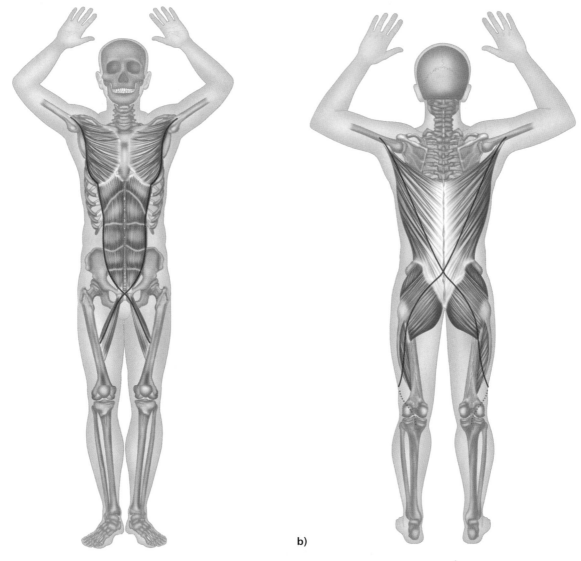

a)   b)

Figure 8.18: The two Functional Lines; a) Front Functional Line, b) Back Functional Line.

Deep Front Line

The Deep Front Line (DFL) forms a complex core volume from the inner arch of the foot, up the inseam of the leg, into the pelvis and up the front of the spine to the bottom of the skull and the jaw. This 'core' line lies between the Front and Back Lines in the sagittal plane, between the two Lateral Lines coronally, and is wrapped circumferentially by the Spiral and Functional Lines. This line contains many of the more obscure supporting muscles of our anatomy, and because of its internal position has the greatest fascial density of any of the lines (*see* figure 8.19).

Structurally, this line has an intimate connection with the arches, the hip joint, lumbar support, and neck balance. Functionally, it connects the ebb and flow of breathing (dictated by the diaphragm) to the rhythm of walking (organized by the psoas). In the trunk, the DFL is intimately linked with the autonomic ganglia, and thus uniquely involved in the sympathetic / parasympathetic balance between our neuro-motor 'chassis' and the ancient organs of cell-support in the ventral cavity.

The importance of the DFL to posture, movement, and attitude cannot be over-emphasized. A dimensional understanding of the DFL is necessary for successful application of nearly any method of manual or movement therapy. Because many of the movement functions of the DFL are redundant to the superficial lines, dysfunction within the DFL can be barely visible in the outset, but these dysfunctions will gradually lead to larger problems. Restoration of proper DFL functioning is by far the best preventive measure for structural and movement therapies.

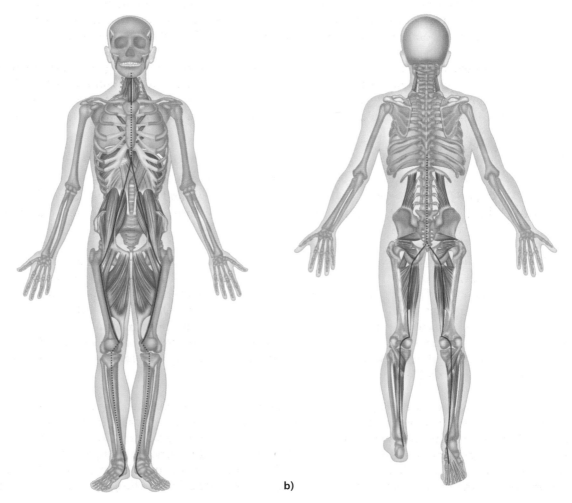

a)   b)

Figure 8.19: The Deep Front Line (DFL); a) anterior view, b) posterior view.

Practical Applications

How does this Anatomy Trains Myofascial Meridian idea add to our practical strategizing for manual therapy?

A look at the Superficial Front Line from the side reveals how useful work on the front of the shin can be to sorting out certain lower back problems and even forward head posture.

Knowing that the plantar fascia and soleus-gastrocnemius complex are joined around the periosteum of the heel allows us to see that in cases where the weight is shifted forward (the ubiquitous 'on your toes' posture), the heel – which should act as the 'kneecap of the ankle' – is instead forced by the tension along the Superficial Back Line into the subtalar joint, limiting movement and reducing support for the back of the body.

Understanding the connection between the lateral longitudinal arch and the hip abductors via the Lateral Line enables the practitioner to make the link between foot balance and pelvic balance, leading to successful soft-tissue strategies for genu varus and valgus. The Spiral Line shows the relation between pelvic tilt and inner arch support, or how to resolve a lateral head shift by working with the opposite shoulder.

Numerous other examples in clinical application are offered in *Anatomy Trains*, and the supporting video programs and courses. Every therapist has seen shoulders drop away from the ears when the feet and legs are worked, low back pain melt away from work in the groin, or a client's breathing open from work on the forearms. The Anatomy Trains map offers one way of understanding and managing these effects in terms of mechanical or energetic communication across our 'sinew channels' of the fascial connections.

Once the relationships within each line are understood, the interactions among the lines open new possibilities for resolving long-standing postural and movement patterns which will not yield to 'single part' attempts to remedy a problem. Progressive work with the lines can create dynamic shifts in these patterns, resulting in the re-introduction of 'poise' – an integral balance and length in body structure.

Larger Considerations

One of the sequelae from the recent industrial and electronic revolutions is a society increasingly alienated from its body. While a few hone their kinesthetic skills through sport and dance (while others hone their reflexes with sophisticated computer games), many more are losing muscle mass, losing an accurate body image, and generally losing 'touch'.

Physical education and manual therapy, in both their traditional and holistic forms, seek to restore balance, awareness, proper functioning, and a healthy relationship with the physical self. New models, such as the concepts outlined above and other systems-oriented views, open new avenues for a populace weakened by constant sitting, fixed focal lengths, improper footwear treading relentlessly flat surfaces, cheapened sexuality, reduced contact with the natural world, lack of activity, and poor education concerning their physical selves from infancy on up. One major challenge for the 21st century is to adapt body systems forged in a Neolithic world to the socially crowded and almost entirely man-made environment we are rapidly constructing worldwide.

We are accustomed to the idea of IQ – measuring the intelligence of the brain. We are becoming more accustomed to EQ – the idea of emotional intelligence. What is needed is a map to the territory of KQ – kinesthetic intelligence, the intelligence of the body in motion. From the skill and awareness that makes an awkward body graceful to the inherent sense that warns us of impending danger; from the precise coordination required in a basketball lay-up to the body memory involved in plucking just the right strings on a harp; from the wisdom of rest and activity cycles to the cellular letting go required to forgive

– there is great intelligence in the body that is not yet well understood. Therefore it is not being taught, and therefore it is being progressively lost, except for small pockets within Eastern and Western medicine where what the great physiologist Walter Cannon called the 'wisdom of the body' is being honored and developed. The most reasonable part in us is the part that does not reason.

These various lines of inquiry into KQ could be gathered under the banner of 'Spatial Medicine' (as opposed to the medicine of Matter [allopathic or nutritional], or the medicine of Time [psychotherapy or shamanism]). What can we learn from how humans are arranged in space, and how they perceive and work with their spatial arrangement? Osteopathy, chiropractic, orthopedics and physiotherapy would qualify as Spatial Medicine. So do the entire alphabet of new (and old) therapies from Alexander, Bioenergetics and Continuum, through Feldenkrais and Gyrotonics, to Rolfing, Somatics, and Tai Chi, all the way to Yoga and Zero Balancing. All these (and the many more not named) are inquiries into our spatial relationships and their meaning, and all seem to contribute to the whole picture. Shifting the positions of bones, altering the length of fascial and myofascial tissues, and training the neuro-muscular system all aim for the same goal – easy, generous, poised movement, structural stability, and the extension of healthy movement into later life.

In short, a systems view (as opposed to the symptoms view) of our structural and movement selves is required to counter the destructive effects of the world we have created for ourselves. The anatomical details so vividly and economically set forth in this book can help with the task of finding, restoring, appreciating, and properly using our amazing locomotor system. So can new overall organizing schemes like the Anatomy Trains – the ever-smaller can be put into service of the ever-larger, and vice versa. True human intelligence – what Norbert Weiner called 'the human use of human beings' – will be attained not by transcending the physical self, but only through our full participation with our marvelous physicality.

Resources

Anderson, D. M. (chief Lexicographer): 2003. *Dorland's Illustrated Medical Dictionary, 30th edition*. Saunders, an imprint of Elsevier, Philadelphia.

Biel, A.: 2001. *Trail Guide to the Body, 2nd edition*. Books of Discovery, Boulder.

Clemente, C. M. (editor): 1985. *Gray's Anatomy of the Human Body, 30th edition*. Lea & Febiger, Philadelphia.

deJong, R. N.: 1967. *The Neurological Examination, 2nd & 3rd editions*. Harper & Row, New York.

Foerster. O., & Bumke, O.: 1936. *Handbuch der Neurologie (vol. V)*. Publisher unknown, Breslau.

Haymaker, W., & Woodhall, B.: 1953. *Peripheral Nerve Injuries, 2nd edition*. W. B. Saunders Co., Philadelphia.

Jarmey, C.: 2004. *The Atlas of Musculo-skeletal Anatomy*. Lotus Publishing, Chichester / North Atlantic Books, Berkeley.

Kendall, F. P., & McCreary, E. K.: 1983. *Muscles, Testing & Function, 3rd edition*. Lippincott, Williams and Wilkins, Baltimore.

Lawrence, M.: 2004. *Complete Guide to Core Stability*. A & C Black, London.

Myers, T. W.: 2001. *Anatomy Trains*. Elsevier, Edinburgh.

Romanes, G. J. (editor): 1972. *Cunningham's Textbook of Anatomy, 11th edition*. Oxford University Press, London.

Schade, J. P.: 1966. *The Peripheral Nervous System*. Elsevier, New York.

Spaleholz, W.: (date unknown). *Hand Atlas of Human Anatomy (vols. II & III, 6th edition)*. J. B. Lippincott, London.

Tortora, G.: 1989. *Principles of Human Anatomy, 5th edition*. Harper & Row, New York.

Index

Index of Muscles